Springer Theses

Recognizing Outstanding Ph.D. Research

Aims and Scope

The series "Springer Theses" brings together a selection of the very best Ph.D. theses from around the world and across the physical sciences. Each volume has been selected for its scientific excellence by an internationally renowned institute or department. For greater accessibility to non-specialists, the published versions include an extended introduction, as well as a foreword by the student's supervisor explaining the special relevance of the work for the field. The series as a whole provides a valuable resource both for newcomers to the research fields, and for other scientists seeking detailed background information on special questions. Finally, it serves as an accredited documentation of the valuable contributions made by today's younger generation of scientists.

How to Apply

Theses prepared at internationally leading universities and institutes may be nominated for publication by supervisors with the endorsement of the head of department. Each such department is entitled to nominate one thesis per year. Enquiries and requests for a nomination form should be sent, together with the thesis, to Ms. Nathalie Berg (see link on right).

Eligibility Criteria

To qualify for consideration, theses should fulfil the following criteria:

- They must be written in good English.
- The topic should fall within the confines of Chemistry, Physics, Earth Sciences, Engineering and related interdisciplinary fields such as Materials, Nanoscience, Chemical Engineering, Complex Systems and Biophysics.
- The work reported in the thesis must represent a significant scientific advance.
- A maximum of 30% of the text may be a verbatim reproduction from the author's previous publications.
- They must have been examined and passed during the 12 months prior to nomination.
- The theses should have a clearly defined structure including an introduction accessible to new PhD students and scientists not expert in the relevant field.
- After acceptance, theses should be supplemented by a "Supervisor's Foreword" outlining the wider scientific relevance of the work.

Indexed by zbMATH.

Eleonora Polini

Broadband Quantum Noise Reduction in Advanced Virgo Plus

From Frequency-Dependent Squeezing Implementation to Detection Losses and Stray Light Mitigation

Springer

Eleonora Polini
Chargée de recherche CNRS, ARTEMIS
Laboratory, Observatoire de la Côte d'Azur
Université Côte d'Azur
Nice, France

ISSN 2190-5053 ISSN 2190-5061 (electronic)
Springer Theses
ISBN 978-3-031-95142-8 ISBN 978-3-031-95143-5 (eBook)
https://doi.org/10.1007/978-3-031-95143-5

This Springer imprint is published by the registered company Springer Nature Switzerland AG
The registered company address is: Gewerbestrasse 11, 6330 Cham, Switzerland

Foreword

Ground-based gravitational waves detectors are regularly upgraded in order to improve their reach and the rate of gravitational waves detections. The first observation of a gravitational wave emitted by a merger of a binary black hole lead to the Physics Nobel Prize in 2017. Then the LIGO-Virgo network enabled the detection of a coalescence of neutron stars, which initiated the so-called multi-messenger astronomy. This has opened up a new way of observing the Universe with the regular detection of astrophysical sources of gravitational waves. Eleonora Polini's thesis is focused on the upgrade of the Virgo detector (Advanced Virgo+) which took place between the O3 and O4 data taking, her work being part of the overall improvements to the network of gravitational wave detectors. The main topic of the PhD is the quantum noise reduction in Advanced Virgo+ with the implementation of the frequency dependent squeezing (FDS) technique. The FDS is a new technique, which had never been implemented on a GW detector and allows a reduction of the quantum noise in all the detector bandwidth. The thesis work also concerns a part of the detection system upgrade closely linked with quantum noise reduction. During her PhD Eleonora Polini worked on the development, the installation, and the commissioning of optical benches for the squeezing injection subsystem, composed of two new in-vacuum suspended benches used to propagate the beams generated by the squeezer to a filter cavity, and then to inject the squeezed states of light into the interferometer. After these benches were installed, the work focused on the onsite commissioning with broad FDS commissioning activities. In this context Eleonora implemented a noise subtraction loop in order to remove the scattered light noise that was spoiling the homodyne photodiodes signal. This achievement was fundamental to allow the measurement of the squeezing level at low frequency. Moreover, this successful implementation demonstrates the feasibility of such a mitigation technique of scattered light that could be useful for other parts of the interferometer or more generally. In order to guarantee a good performance of the squeezing technique, the losses have to be minimized on the path of the interferometer output beam. Eleonora Polini also contributed to this aspect with the characterization of the new (high finesse) output mode cleaner cavity (OMC) which is filtering the interferometer output beam. This work was followed by the installation of the OMC on site and its commissioning and includes also the optimization of the mode matching of this cavity with the interferometer beam, which was brought to the level of 99%. Another topic of the PhD is the characterization and reduction of scattered light on the Virgo optical benches, as this can lead to the deterioration of the interferometer's sensitivity. First, Eleonora Polini worked on the reduction of the scattered light due to ghost beams on a large fraction of the optical benches of Virgo. This work was completed with the installation of beam dumps and diaphragms on the benches. Second, she also participated in tabletop experiment at LAPP to measure the light scattered by optical elements at small angle. All these contributions are very well detailed and explained in the thesis with sufficient introduction to each topic. The quality of Eleonora's work was recognized by the Virgo award prize, and the Gravitational Wave International Committee (GWIC) recognized the importance of her work to the field of gravitational waves with the 2022 GWIC-Braccini thesis prize. It was a pleasure for me to accompany Eleonora Polini's work through this variety of important topics for gravitational wave detectors.

Annecy, France
April 2025

Dr. Edwige Tournefier

Preface

Gravitational astronomy began with the first direct detection of gravitational waves, emitted by the merger of black holes, by the LIGO detectors in the United States. This breakthrough was followed by the addition of the European Virgo detector to the observation network, which enabled more precise localization of gravitational wave sources. A gravitational signal from the coalescence of neutron stars was detected, which produced an electromagnetic signal that was detectable with other optical telescopes, thus expanding the so-called multi-messenger astronomy and paving the way for astrophysical discoveries. By the end of the O3 observing run, 90 coalescence events had been detected, with an event rate of approximately one per week. The detectors were further improved in preparation for the next O4 observing run, allowing for the observation of an even larger portion of the Universe. Among the sources of noise limiting the sensitivity of current gravitational wave interferometric detectors, one finds quantum noise, caused by the recombination of quantum vacuum fluctuations with the detector light. This noise is present both at high frequencies in the form of shot noise and at low frequencies as radiation pressure noise. To reduce it, a quantum technique was implemented during the O3 observation run, involving the introduction of special quantum states called vacuum squeezed states. This led to a reduction of quantum noise at high frequencies but also caused a slight increase in noise at low frequencies. To achieve broadband quantum noise reduction, a more complex system that rotates squeezed states at the appropriate frequency is required. The implementation of the frequency-dependent squeezing system in Virgo is thoroughly described in this work. Additionally, the recombination of stray light, i.e., back-reflected or back-scattered light, with the interferometer light can spoil the low frequency sensitivity of Virgo and also obscure the benefits introduced by the quantum noise reduction system. This work discusses simulations and measurements aimed at studying and mitigating this noise source in Virgo. Finally, since squeezed vacuum states are very sensitive to degradation due to optical losses, actions have been taken to reduce losses at the most critical points of the detector. In particular, this work discusses modifications made to the detection system that allowed for a substantial reduction in optical losses, thus enabling even greater quantum noise reduction.

Part I includes the introductory chapters on gravitational waves. Chapter 1 presents the theoretical derivation of the gravitational wave signal, listing all possible sources and detailing all events detected by the end of the O3 observation run. Chapter 2 explains the principle of gravitational wave detection through laser interferometry and describes all noise sources that degrade its sensitivity.

Part II includes chapters on broadband quantum noise reduction in Virgo. Chapter 3 presents a theoretical description of quantum noise and how squeezed vacuum states can reduce it. Chapter 4 describes the system used to produce frequency-dependent squeezing in Virgo. Chapter 5 details the experimental work to commission this system, culminating in the first measurement of squeezing state rotation as a function of frequency.

Part III includes chapters on stray light noise. Chapter 6 explains the mechanism of stray light recombination with the interferometer and shows simulations of the noise coming from the different suspended benches of Virgo for O4. Chapter 7 discusses the study and mitigation of ghost beams on all of Virgo's benches. Chapter 8 describes a system developed to control scattered light in the squeezing system.

Part IV includes chapters related to the reduction of optical losses in the detection system, which interfaces the squeezing system with the interferometer. Chapter 9 describes the work done regarding the injection of squeezing into Virgo. Chapter 10 presents the work on replacing the output mode cleaner cavities with a single cavity and other important upgrades aimed at reducing optical losses in squeezed states.

Part V provides an outlook on the work carried out.

Keywords Gravitational waves detection, Advanced Virgo Plus, Quantum noise, Optical cavities, Suspended Benches, Frequency-dependent squeezing, Scattered light, Output mode cleaner cavity

Nice, France
March 2025

Eleonora Polini

Acknowledgments

I began this thesis a few months before the onset of the pandemic, and despite the many challenges that followed, I will always remember these years as profoundly enriching, both personally and professionally. I am deeply grateful to all the people who have contributed to my growth over this time—those who have helped, supported, and encouraged me along the way. My heartfelt thanks go to each of you. Without your presence and guidance, these years would not have been as fulfilling, and I would not carry such a wide smile when I look back on them.

I would like to begin by thanking Lisa Barsotti and Matteo Barsuglia for agreeing to review my thesis. Thank you for your careful reading, valuable comments, and insightful suggestions.

My sincere thanks to Fiodor Sorrentino, Laurent Derome, and Stephane Jezequel for examining my work.

A special thanks to PhD advisor Edwige Tournefier. Thanks for your meticulous supervision in editing this manuscript, and for the many rounds of corrections! Thank you for the stimulating scientific discussions throughout these years of my PhD.

I am also grateful to Romain Bonnand and Romain Gouaty for reading my thesis and for sharing their extensive knowledge of optics and suspended benches. Thank you for your patience in teaching me the role of an experimental physicist—from climbing Virgo towers to setting up optical benches.

Thanks to Michal Was for sharing your knowledge on scattered light and output mode cleaner with me. Your unwavering patience in answering my persistent questions about various aspects of the detector was invaluable; I learned so much from you!

Thanks to Marco Vardaro for supervising my work during my time at Virgo and for introducing me to the role of "commissioner," always with great patience and passion. I am also thankful for all the continued help and support you've provided.

Thanks to Raffaele Flaminio for opening the doors of National Astronomical Observatory of Japan (NAOJ) and the Laboratoire d'Annecy de Physique des Particules (LAPP) to me, and for allowing me to pursue the topics I am most passionate about.

Thanks to Alain, Loic, Bruno, Laurent, Fabrice, and the other members of the group at LAPP for their continuous support during my work.

I would also like to express my gratitude to all the members of the Virgo collaboration who have helped me over the years in understanding various aspects of the detector.

Special thanks to the squeezing commissioning team—Alessandro, Andrea, Barbara, Eleonora, Fiodor, Henning, Jean-Pierre, Marco, Martina, Matteusz, Moritz, Matteo, Valeria, Yuefan—for trusting me and patiently teaching me all aspects of this fascinating and complex experiment.

Thanks to all the researchers who always answered my questions on the most varied topics in the control room, from Antonino, to Diego, Maddalena, Manuel, Julia, Paolo... and all the others!

Thanks also to the electronic and mechanical teams and the Virgo vacuum group for their support even if asked with short notice.

Thanks to Ettore for always being my point of reference and support, advising and encouraging me even though I am no longer in Rome. This has meant so much to me.

I would like to thank all my friends who have added so much color to these three years in so many different ways. Despite the pandemic, the distance, and countless difficulties, you've always made me feel at home. Many thanks to each of you. Many thanks to each of you.

A special thanks to Lella, *la mia compagna di vita*, for so many years that I almost lost count. It goes without saying that your warm support has been fundamental for my entire career and, by extension, to this thesis.

Thanks to Colomba, who despite her incessant misfortune in recent years has always found a way to be close to me even if far away.

Thanks to Alessandro and Letizia, with whom I had the pleasure of experiencing Annecy under the snow, both in April and December!

I am grateful to my colleagues, and now friends, at LAPP: Alexandre, Anthony, Camille, Cervane, Christopher, Felipe, Gitu, Iuri, Kike, Leo, Luka, Matteo, Michel, Monica, Paul, Quim, and Victor. I never imagined finding such incredible people to share lab lunches, amazing chalet weekends, and much more with.

Thanks to the friends of the Soirée group: Aga, Alessandro, Ben, Daniela, Diana, Edwin, Eleonora, Ilenia, Leonardo, Manon, Marta, Martina, Matthieu, Nathan, Roberta, Sergio, Tam, and Virginie. I never thought I would meet people as lively and full of *joie de vivre* as you. Thank you for all the activities, aperitifs, nights out, and, especially, the laughter.

Thanks to the fantastic colleagues and friends who have been with me during my year in Virgo: Agnese, Andrea, Diego, Eleonora, Enzo, Julia, Julien, Lorenzo, Manuel, Marco, Mattia, Matthieu, Michele, Riccardo, Yuefan ... Thanks for all the *amari al SUD* over these years and for all the support you have given me. You made me feel part of a family more than a collaboration, despite all the daily problems we encountered.

Thanks to my lifelong friends in Rome, who always welcome me with open arms, as if I had never left: Flavia, Valerio, Adriana, Simonetta, Giada, Beatrice, Caterina, Camilla, Gianluca . . .

Lastly, thanks to my family, who continues to support me even from a distance.

Competing Interests The author has no competing interests to declare that are relevant to the content of this manuscript.

Contents

Acronyms

AdV	Advanced Virgo
AdV+	Advanced Virgo Plus
AEI	Albert Einstein Institute, Hannover, Germany
AR	Anti-Reflective
AOM	Acousto-Optic Modulator
BAB	Bright Alignment Beam
BBH	Binary Black Holes
BNS	Binary Neutron Stars
BRDF	Bidirectional Reflectance Distribution Function
BS	Beam Splitter
CC	Coherent Control
CMRR	Common Mode Rejection Ratio
CP	Compensation Plate
EOM	Electro-Optic Modulator
EQB1	External sQueezing Bench 1
EQB2	External sQueezing Bench 2
ET	Einstein Telescope
DC	Direct Current
FDS	Frequency-Dependent Squeezing
FIS	Frequency-Independent Squeezing
FC	Filter Cavity
FCEM	Filter Cavity End Mirror
FCIM	Filter Cavity Input Mirror
FI	Faraday Isolator
FP	Fabry-Perot
FWHM	Full Width at Half Maximum
GR	GReen
GW	Gravitational Wave
HD	HomoDyne
HG	Hermite-Gaussian
HR	Highly Reflective

HWP	Half-Wave Plate
IMC	Input Mode Cleaner
IR	InfraRed
ITF	InTerFerometer
KAGRA	KAmioka GRAvitational wave detector
LAPP	Laboratoire d'Annecy de Physique des Particules, France
LG	Laguerre-Gaussian
LIGO	Laser Interferometric Gravitational wave Observatories, USA
LMA	Laboratoire des Materiaux Avances, France
LO	Local Oscillator
LVDT	Linear Variable Differential Transformer
MZ	Mach-Zehnder
NAOJ	National Astromical Observatory of Japan
NE	North End mirror
NI	North Input mirror
NSBH	Neutron Star—Black Hole binaries
OMC	Output Mode Cleaner
OPO	Optical Parametric Oscillator
PD	PhotoDiode
PDH	Pound-Drever Hall
PLL	Phase-Locked Loop
POP	Pick-Off Plate
PPKT	Periodically-Poled potassium (K) Titanyl Phosphate
PR	Power Recycling
PRC	Power Recycling Cavity
PSD	Position Sensor Device
PTFE	PolyTetraFluoroEthylene
PZT	lead (Pb) Zirconate Titanate
QD	Quandrant Photodiode
RF	Radio-Frequency
RMS	Root Mean Square
RoC	Radius of Curvature
RTL	Round-Trip Losses
SC	Sub-Carrier
SDB1	Suspended Detection Bench 1
SDB2	Suspended Detection Bench 2
SHG	Second Harmonic Generator
SNR	Signal-to-Noise Ratio
SPRB	Suspended Power Recycling Bench
SQB1	Suspended sQueezing Bench 1
SQB2	Suspended sQueezing Bench 2
SLL	Stray Light Loop
SR	Signal Recycling
SRC	Signal Recycling Cavity
TEM	Transverse ElectroMagnetic

TF	Transfer Function
TGG	Terbium Gallium Garnet
TIS	Total Integrating Scattering
WE	West End mirror
WI	West Input mirror

Part I
Gravitational Waves Detection

Chapter 1
Gravitational Waves

Gravitational waves (GWs) are 'ripples' in the fabric of space-time caused by the most energetic and violent phenomena in the Universe. They were predicted by Albert Einstein in 1916 in his theory of general relativity [1]. In this Chapter, it is described the derivation of the gravitational waves equation under the linear approximation and the sources that can generate them are illustrated.

1.1 Gravitational Waves Derivation from Einstein Equation

The equation describing the gravitational waves can be derived from the general relativity equations of Albert Einstein [2]. This equation relates the space-time metric to the energy content:

$$G_{\mu\nu} = R_{\mu\nu} - \frac{1}{2}g_{\mu\nu}R = \frac{8\pi G}{c^4}T_{\mu\nu} \qquad \mu, \nu = 1, 2, 3, 4 \tag{1.1}$$

where $G_{\mu\nu}$ is called the Einstein tensor, $R_{\mu\nu}$ is the Ricci curvature tensor, R is the Ricci curvature scalar, $g_{\mu\nu}$ is the metric, c is the speed of light, G is the Newtonian constant of gravitation and $T_{\mu\nu}$ is the stress-energy tensor.

The expansion of these equations is given in the approximation of being around the flat-space metric:

$$g_{\mu\nu} = \eta_{\mu\nu} + h_{\mu\nu} \tag{1.2}$$

where $\eta_{\mu\nu} = diag(-1, 1, 1, 1)$ is the flat Minkowski metric and $|h_{\mu\nu}| << 1$ is the small perturbation term.

In this discussion the background curvature of the space-time that is crucial for the large scale effects, will be neglected. In other words, it is highlighted

E. Polini, *Broadband Quantum Noise Reduction in Advanced Virgo Plus*,
Springer Theses, https://doi.org/10.1007/978-3-031-95143-5_1

the existence of small perturbations of the metric that propagate in space and time superimposed on the almost static curvature due to the astrophysical objects. However there will be considered only the wave solutions characterized by much shorter wavelengths than the typical scale of background curvature. In many cases, the deviation from the flat metric induced by the presence of a matter-energy distribution is small enough to justify this approximation. By choosing an opportune set of coordinates and in absence of sources ($T_{\mu\nu} = 0$, propagation in empty space) the field equation 1.1 takes the form of a classical wave equation [2]

$$\Box_F h_{\mu\nu} = \left(-\frac{\partial^2}{c^2\partial t^2} + \nabla^2\right) h_{\mu\nu} = 0 \tag{1.3}$$

where $\Box_F$ is D'Alambert operator in flat space. The general solution is a wave propagating at the speed of light

$$h_{\mu\nu}(\mathbf{x}, t) = \epsilon_{\mu\nu} \mathrm{e}^{i(2\pi f t - \mathbf{k}\cdot\mathbf{x})} \tag{1.4}$$

where f is the gravitational wave frequency, $\epsilon_{\mu\nu}$ is the polarization tensor and $\mathbf{k}$ is the wave vector.

Defining an opportune gauge, the *transverse-traceless gauge* or TT gauge, the metric for a wave propagating along the x direction can be written as

$$h_{\mu\nu}^{TT} = \begin{pmatrix} 0 & 0 & 0 & 0 \\ 0 & 0 & 0 & 0 \\ 0 & 0 & h_+ & h_\times \\ 0 & 0 & h_\times & -h_+ \end{pmatrix} = \begin{pmatrix} 0 & 0 & 0 & 0 \\ 0 & 0 & 0 & 0 \\ 0 & 0 & h_{yy} & h_{yz} \\ 0 & 0 & h_{yz} & -h_{yy} \end{pmatrix} \tag{1.5}$$

where h_+ and $h_\times$ are the amplitudes of the *plus* and *cross* polarization of the wave. In the TT gauge the components of the metric tensor $h_{\mu\nu}$ are different from zero only on the plane orthogonal to the direction of propagation (transverse), and $h_{\mu\nu}$ is traceless. In conclusion, a gravitational wave has only two physical degrees of freedom which correspond to the two possible polarization states.

The wave can be considered as a linear combination of two different components

$$h_+ = \epsilon_+ \mathrm{e}^{i(2\pi f t - kx)} \tag{1.6}$$

and

$$h_\times = \epsilon_\times \mathrm{e}^{i(2\pi f t - kx)} \tag{1.7}$$

corresponding to two independent polarization states. The effect of the two polarizations on a ring of freely falling particles is shown in Fig. 1.1.

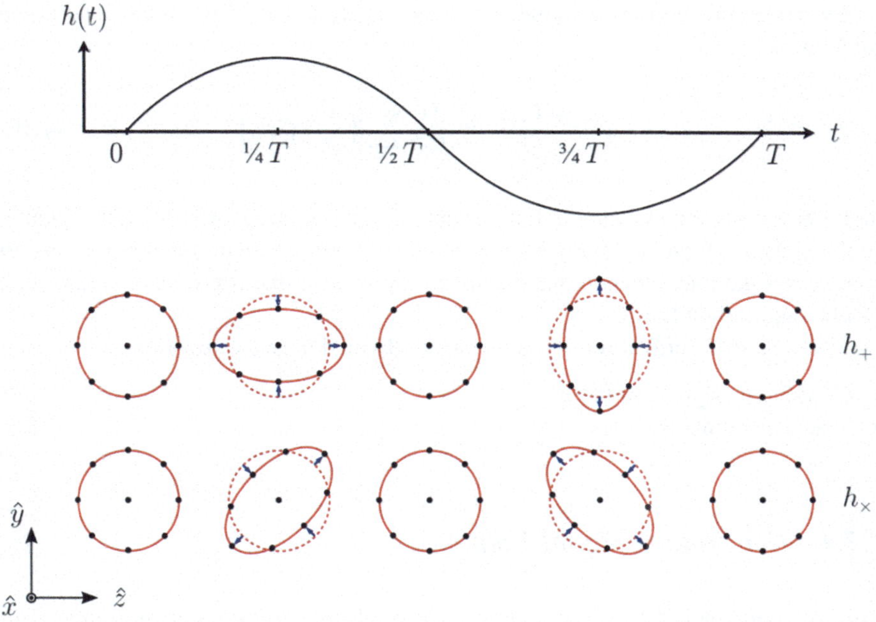

Fig. 1.1 A monochromatic gravitational wave of pulsation $\omega = 2\pi/T$ propagates along the x direction. The lower panel shows the effects of the + and × polarizations on a ring of freely falling particles, in a local inertial frame. Figure adapted from [3]

1.2 Gravitational Waves Sources

Gravitational waves are generated by massive objects in acceleration. To describe how they are generated, the multipolar expansion of the stress-energy tensor $T_{\nu\mu}$ [4] can be used. From this expansion, a convenient choice is to express the spatial components of the wave in the TT gauge [2] as

$$h_{\mu\nu}^{TT}(t) = \frac{2G}{rc^4}\ddot{Q}_{\mu\nu}(t - \frac{r}{c}) \tag{1.8}$$

where $\ddot{Q}_{\mu\nu}$ is the second time derivative of the quadrupole moment of the mass distribution of the source, G is the Newtonian constant of gravitation, c is the speed of light and r is the distance between the source and the observer.

The traceless mass quadrupole moment of the system $Q_{\mu\nu}$, with $\mu, \nu = 1, 2, 3$ is defined as a function of the source mass density $\rho(\mathbf{x}, t)$

$$Q_{\mu\nu} = \int_V \rho \left(x_\mu x_\nu - \frac{1}{3}\delta_{\mu\nu}x^2 \right) d^3x \tag{1.9}$$

where $\delta_{\mu\nu}$ is the Kronecker delta.

Moreover, the luminosity radiated by a moving system in terms of the quadrupole moment is

$$L_{GW} = \frac{dE_{GW}}{dt} = \frac{G}{5c^5} \sum_{kh} \left(\frac{d^3}{dt^3} Q_{kh} \right)^2 \tag{1.10}$$

and it expresses the rate at which the energy E_{GW} is carried away from the source. In the Universe there are many sources that can produce gravitational radiations. In some cases the phenomenon that occurs is so powerful that if produces signals that can be detected on Earth.

The sources of gravitational waves can be divided in two categories:

- Continuous signal emitters;
- Transient signal emitters.

1.2.1 Continuous Signal Emitters

This category includes all the sources that produce a quasi-monochromatic with a quasi-constant amplitude signal. Examples of continuous sources are binary star systems which rotate one around each other and compact stars that rotate rapidly around one of their axes, with a certain degree of asymmetry. Stochastic gravitational background signals can also be included in this category and they represent the gravitational analogue of the cosmic microwave background electromagnetic radiation.

1.2.1.1 Rotating Neutron Stars

A rotating neutron star with an asymmetry on its surface can emit gravitational radiation [5]. A subset of the expected population of neutron stars in the Galaxy has been observed as pulsating electromagnetic sources, commonly called *Pulsars*. They emit pulses of radiation at very regular intervals that typically range from milliseconds to seconds [6]. Pulsars have very high magnetic field ($10^{11} - 10^{15}$ Gauss) which direct jets of particles out along the two magnetic poles. Often the rotation axis does not coincide with the magnetic axis and the star emits electromagnetic waves along the direction of the magnetic field. An observer on Earth intercepts the beam a regular time intervals and then receives radio pulses with a period of recurrence equal to that of rotation of the star. In order to be a source of gravitational waves they have to show a certain degree of asymmetry, to allow the quadrupole moment to vary with time, as shown in Eq. 1.8. The expected gravitational wave amplitude [7, 8] associated to a source at distance r from the observer and rotating at the frequency f_{rot} around its symmetry axis z, is

$$h = \frac{4\pi^2 G\epsilon I_{zz} f_{GW}^2}{c^4 r} = 1.06 \times 10^{-26} \left[\frac{\epsilon}{10^{-6}}\right]\left[\frac{I_{zz}}{I_0}\right]\left[\frac{f_{GW}}{100\text{Hz}}\right]^2\left[\frac{1\text{kpc}}{r}\right] \quad (1.11)$$

where $I_0 = 10^{38}\,\text{kg}\cdot\text{m}^2$ is a nominal neutron star moment of inertia, I_{zz} is the moment of inertia along z, ϵ is the equatorial ellipticity (which takes into account the asymmetry of the system) and $f_{GW} = 2f_{rot}$ is the frequency of the emitted gravitational radiation. Analyzing observation data and for a frequency band of search of 20–2000 Hz, the limit on the amplitude of these signals is set to $h \sim 1 - 2 \cdot 10^{-25}$ [7, 8].

1.2.1.2 Stochastic Background

The stochastic gravitational background is composed of the combination of weak, independent, unresolved gravitational signals from different sources [5]. The main contribution is thought to come from astrophysical sources such as the superposition of signals from black holes and neutron star binaries, supernovae or depleting boson clouds around black holes. Another contribution could come from signals produced by cosmological sources, i.e. produced during the inflationary epoch in the early universe or from the coalescence of primordial black holes. The difference between the cosmological stochastic background and the astrophysical one is that the former is expected to be stationary, Gaussian, non-polarized and isotropic, while the latter cannot be so, since the distribution of galaxies with a distance of up to 100 Mpc is not isotropic. The upper limit for this kind of sources has been set to be $h \sim 9 \times 10^{-24} - 10^{-25}$ [9, 10].

1.2.2 Transient Signal Emitters

This category contains all the sources that produce non-stationary signals, whose amplitude varies over time. Examples of this type of source are the coalescence of binary systems of astrophysical objects. Supernovae, spectacular stellar explosions, also fall into this category as they can produce gravitational wave signals.

1.2.2.1 Compact Binary Coalescence

The coalescence of binary astrophysical systems is currently the only source of gravitational waves that has been observed experimentally. The binary systems that can merge and generate gravitational radiation are:

- binary black holes (BBH);
- binary neutron stars (BNS);
- neutron star and black hole (NSBH).

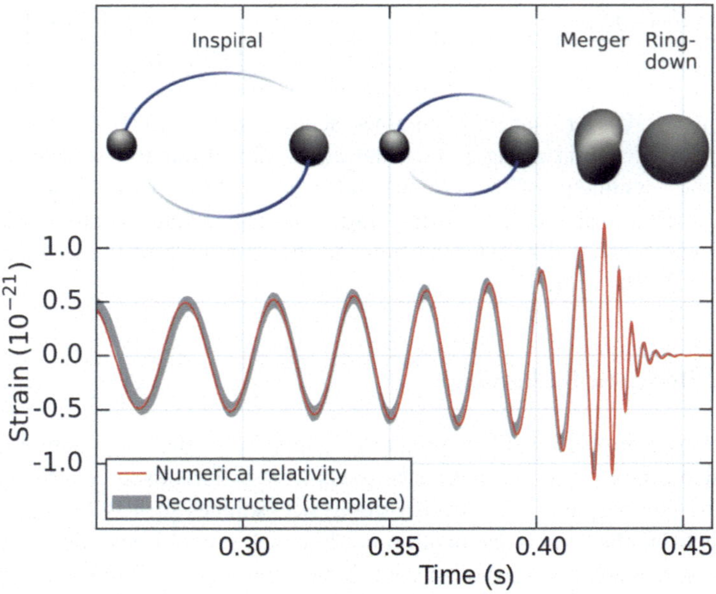

Fig. 1.2 Estimated gravitational wave amplitude from the first event detected by LIGO detectors, named GW150914. The signal is produced from a coalescence of two black holes. The three phases of the merger (inspiral, merger and ringdown) are shown associated to the time evolution of the signal [12]

The two objects rotate around each other, losing energy in the form of gravitational waves. This decreases the distance between the two objects and increases the rotation frequency of the system. This leads to a reduction in the orbital radius of the binary system until the two objects collide one on each other, creating a single black hole. Coalescence occurs in three phases: inspiral, merger and ringdown (see Fig. 1.2). The effective mass of the system during the inspiral phase is called *chirp mass* and it is defined as

$$\mathcal{M} = \mu^{3/5} M^{2/5} = \frac{(m_1 m_2)^{3/5}}{(m_1 + m_2)^{1/5}} \tag{1.12}$$

where m_1 and m_2 are the masses of the two objects of the binary system, $M = m_1 + m_2$ is the sum of the two masses and $\mu = \dfrac{m_1 m_2}{m_1 + m_2}$ is the reduced mass. In this part of the signal both the amplitude and the frequency of the emission are increasing, as the chirp of a singing bird, see Fig. 1.2. The amplitude of the gravitational signal [11], at a distance r from the source, is

$$h(t) = \frac{4\pi^{2/3} G^{5/3} \mathcal{M}^{5/3} \nu(t)^{2/3}}{c^4 r}$$
$$\simeq 1.02 \times 10^{-23} \left[\frac{\mathcal{M}}{M_\odot}\right]^{5/3} \left[\frac{\nu(t)}{100\text{Hz}}\right]^{2/3} \left[\frac{100\text{Mpc}}{r}\right] \tag{1.13}$$

where $\nu(t)$ is the gravitational radiation frequency.

In the past years, 90 gravitational wave events of this category has been observed from LIGO and Virgo detectors (see Sect. 1.3).

1.2.2.2 Supernovae

Core collapse supernova (CCSN) explosion is an other possible source of gravitational waves and it represents one of the most spectacular events in the Universe.

Supernovae with a mass in the range $8M_\odot \leq M \leq 130M_\odot$ can release gravitational radiation during the core collapse of the massive star [13]. Most of the energy during this process is stored as heat in the proto-neutron star and then almost entirely released as neutrinos, with a small fraction being released as kinetic energy during the explosion or emitted as electromagnetic radiation. An even smaller fraction of energy can be released as gravitational waves. Gravitational radiation is however emitted if the core collapse supernova explosion shows an asymmetry. The spherical symmetry of the CCSNs system can be easily altered by the rotation of the star, convective motions in the proto-neutron star and standing accretion shock instability. The amplitude of the gravitational waves generated by this source varies depending on the deviation from spherical symmetry. For an explosion occurring at a distance of 10 kpc from the observer, the amplitude of the gravitational radiation has been estimated to be of the order of $10^{-23} - 10^{-20}$ and the signal duration to vary between 1 ms and few seconds with a frequency of 1 – few kHz [13]. However, these signals have not yet been observed.

1.3 Gravitational Waves Detections

The first gravitational wave signal from a coalescence of two black holes was detected on 14 September 2015 (GW150914) by the LIGO detectors in the USA [12]. Since then, a new era of observing the universe has begun. The detector Virgo, in Europe, had joined the detector network on 1 August 2017, leading to the first three-detector detection on 14 August 2017 (GW170814) [14]. On 17 August 2017 (GW170817), gravitational radiation from the coalescence of two neutron stars was detected from the gravitational wave detectors [15], in coincidence with other telescopes, such as Fermi [16], which revealed its electromagnetic counterpart. This observation started the so-called *multi-messenger astronomy*. The third observing

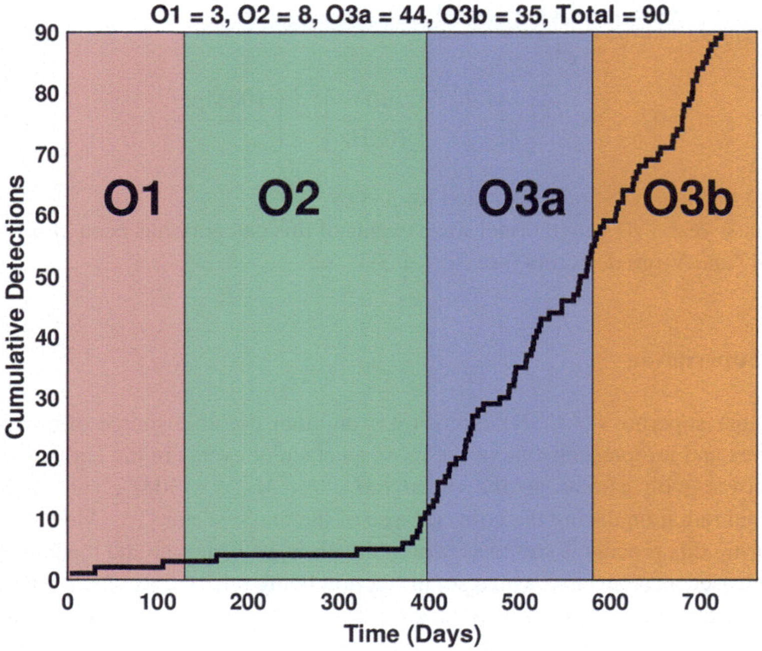

Fig. 1.3 Cumulative count of events and (non-retracted) alerts during O1, O2, O3a and O3b observing runs (Credit: LIGO-Virgo-KAGRA Collaborations)

run led to the detection of several coalescence events, also leading to the detection of two mixed coalescences of a black hole with a neutron star in January 2020 (GW200105 and GW200115) [17]. In total, 85 events has been detected from 2015 until March 2020, when the third observation run ended, see Fig. 1.3: 3 events during O1 [18]; 8 events during O2 [18]; 39 events during O3a [19, 20]; 35 events during O3b [21].

The Fig. 1.4 shows the graphic of the masses of all the gravitational wave events announced from the first detection (O1) until the last events detected (O3b).

1.3.1 Observation Prospects for O4 and O5

After run O3, the commissioning phase of the detectors began in order to perform upgrades that will allow a better sensitivity during the next observation runs. The timeline of the upgrades of LIGO, Virgo and KAGRA are shown in Fig. 1.5. For each observing run it is indicated the BNS range in Mpc as a figure of merit, which quantifies the distance of a detectable BNS system. The run O4 is planned to start in spring 2023 and the target sensitivity for LIGO is 160–190 Mpc, for Virgo is 80–

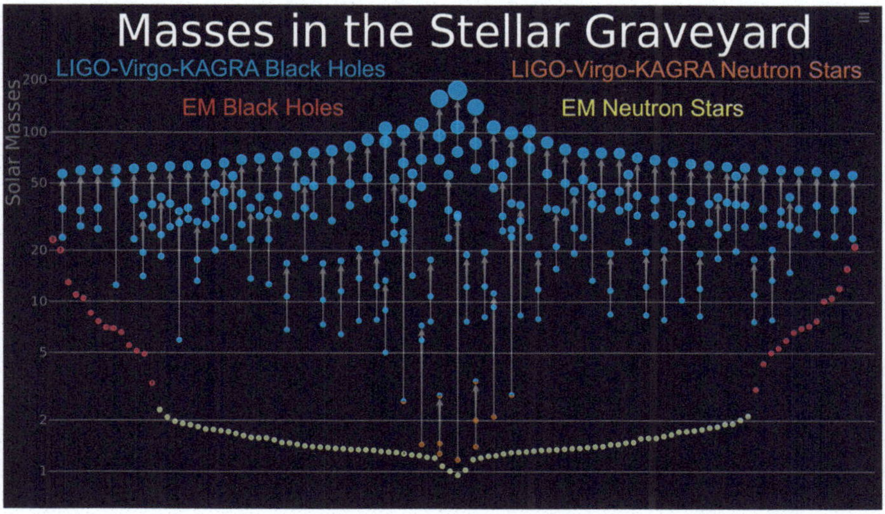

Fig. 1.4 Scheme of the masses of all the gravitational wave events detected from O1 to O3b. The masses of objects detected by gravitational wave signals (black holes in blue and neutron stars in orange) and by electromagnetic radiation (black holes in red and neutron stars in yellow) are shown on the vertical axis. Each vertical arrow represents the coalesce of a binary system in which the lowest-mass object coalesces with the second one on the arrow, creating a higher-mass object where the arrow ends (Credit: LIGO-Virgo/Aaron Geller/Northwestern University)

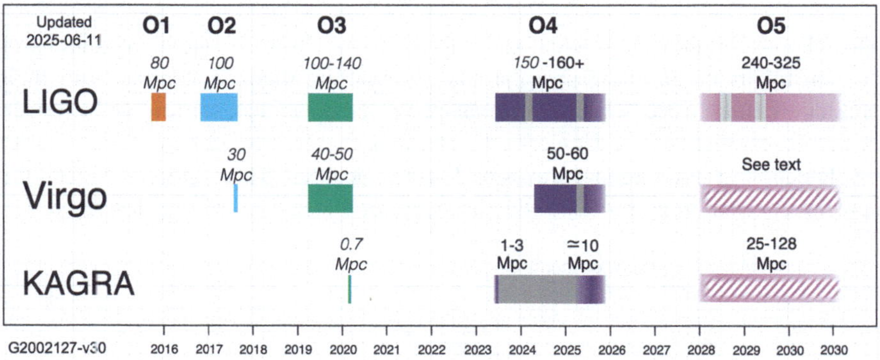

Fig. 1.5 Sensitivity of the detectors LIGO, Virgo and KAGRA as a function of time (Credit: LIGO-Virgo-KAGRA Collaborations)

115 Mpc and for KAGRA is 1–10 Mpc. The duration of the run is planned to be 12 months. Afterwards there will be a new commissioning phase for the detectors in which the sensitivity will be further improved to 240–325 Mpc for LIGO, 150–260 Mpc for Virgo and 25–128 Mpc for KAGRA.

The expected number of detections during O4 using the four gravitational wave detectors and with 90% credibility are [22]

- from 0 to 62 BNS detections with a median 90% credible region for the localization area of 33 deg^2;
- from 0 to 92 NSBH detections with a median 90% credible region for the localization area of 50 deg^2;
- from 35 to 168 BBH detections with a median 90% credible region for the localization area of 41 deg^2.

The large number of binary system detections expected in O4 will allow more precise measurements of the parameters of neutron stars and/or black holes, to characterize the astrophysical distribution of compact binaries in the universe and to test general relativity with these systems [23]. Furthermore, the detection of neutron star coalescences enables multi-messenger astronomy and would allow the study of matter under extreme conditions, i.e. the neutron star equation of state. The compact binaries detections will be of key importance also to have a more precise determination of the Hubble constant. However, during O4 it might be possible to see other gravitational signals not yet detected by LIGO, Virgo and KAGRA such as bursts, continuous waves (both from unknown sources, from pulsars or nearby post-merger neutron stars) and the stochastic background [23].

1.4 Conclusion

This chapter introduces gravitational waves, derived from Einstein's equations of general relativity, and describes the sources emitting them. A century after their prediction, these extremely small signals have been detected, starting a new epoch in the observation of the Universe. In the next chapter, the principle that lead to the detection of gravitational waves on Earth in detailed. The improvements on the detectors needed to go further in the discovery of the cosmos will be described.

Chapter 2
Ground Based Gravitational Waves Detectors

An extremely sensitive instrument is needed to detect gravitational waves. In the early 1960s, the physicist Joseph Weber was the first one who tried to directly detect gravitational waves, after demonstrating that a gravitational radiation can introduce vibration on an harmonic oscillator [24]. After several tentative experiments, it was realized that the best detector on Earth would be a Michelson interferometer with km-scale arms. Indeed, a laser interferometer can measure a change in the distance between two free-falling masses suspended several kilometers apart, 1000 times smaller than the size of a proton. In this chapter, the detection principle of an interferometric gravitational wave detector is described, pointing out the main techniques used to improve the sensitivity with respect to the basic configuration. Moreover, the main sources of noise disturbing the detection of gravitational waves are presented. Finally, the current configuration of the European gravitational wave detector Virgo is described.

2.1 Interferometric Detection Principle

One way to directly observe the amplitude of a gravitational wave is to measure the change in distance introduced between two free-falling masses that the passage of the wave causes.

L represents the distance between the two free test masses and δL the variation in length induced by a perturbation. The amplitude of the gravitational wave, expressed in dimensionless units, is called *strain* or $h(t)$ and it is defined as

$$h(t) = 2\frac{\delta L(t)}{L} \tag{2.1}$$

where the factor 2 indicates that the light is passing back and forth between the two masses.

E. Polini, *Broadband Quantum Noise Reduction in Advanced Virgo Plus*,
Springer Theses, https://doi.org/10.1007/978-3-031-95143-5_2

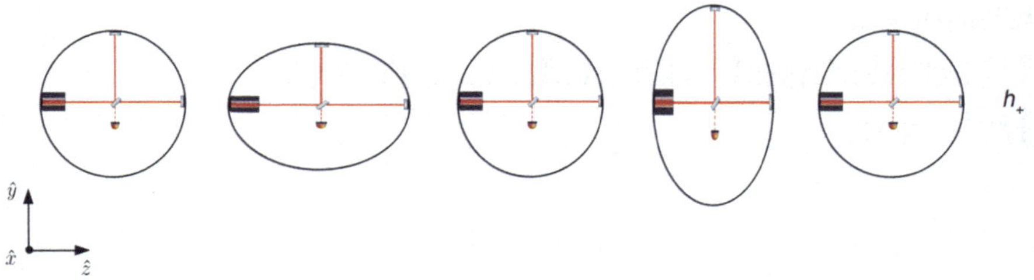

Fig. 2.1 The effect of a passing gravitational wave polarized h_+ which arrives perpendicular to the interferometer plane

The most sensitive ground-based instrument to detect gravitational waves is a km-scale Michelson interferometer. The mirrors that form this detector are suspended, and therefore decoupled from the ground motion, and behave as free masses in the horizontal plane. Figure 2.1 shows the effect of the passage of a polarized h_+ gravitational wave propagating perpendicularly with respect to the interferometer plane.

The arrival of a gravitational wave introduces a differential change in the length of the Michelson arms, which corresponds to the $\delta L(t)$ of Eq. 2.1. The photodetector at the antisymmetric port detects the output power which is proportional to the difference in length between the arms.

2.1.1 *Michelson Interferometer*

The simplest configuration of a Michelson gravitational wave detector is shown in Fig. 2.2.[1]

A beam is generated by a coherent light source and split in two by a 50/50 beam splitter. The two resulting beams propagate perpendicularly along the two arms of the interferometer until they reach the two final mirrors that are denoted with NE (North End) and WE (West End). The arm along the north direction has length L_N while the one in the west direction L_W. The two arms have the same length $L_N = L_W$, which is 3 km in the case of Virgo. The beams recombine in the beam splitter with constructive or destructive interference depending on how the interferometer mirrors are positioned. The chosen configuration is the dark fringe, i.e. destructive interference, to be more sensitive to the gravitational wave signal, as it will be explained in this Section. In the following the relation between the detected output power and the difference in phase of the two beams recombined in the beam splitter is derived.

[1] All the optical schemes presented in this work are based on the Optical Component Library created by A. Franzen [25].

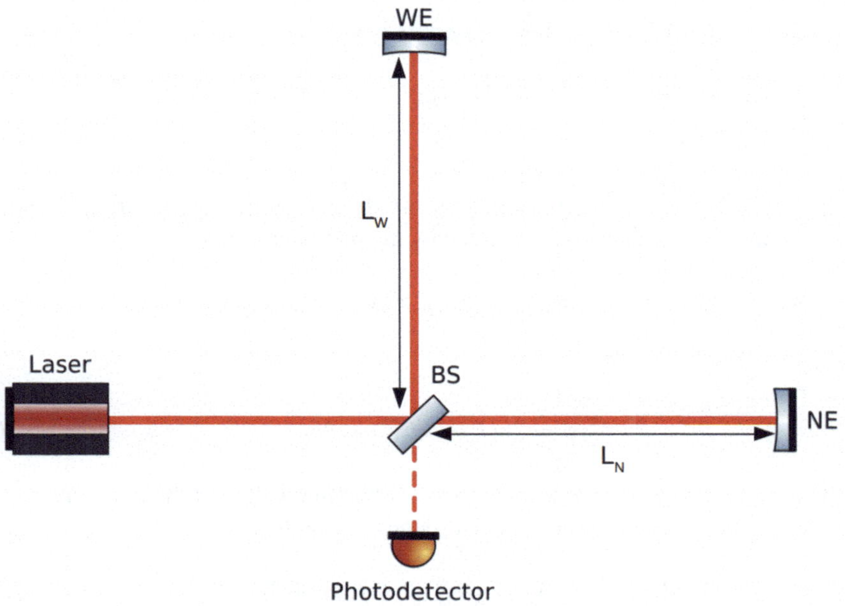

Fig. 2.2 Michelson interferometer scheme. The beam of a coherent light source is split at a 50/50 beam splitter (BS). The two beams travel along interferometer arms of length $L_W = L_N$ and then they are reflected by two end mirrors: the West End (WE) and the North End (NE). The two beams interfere constructively or destructively based on their phase relation. The photodetector is at the antisymmetric port of the interferometer

A laser beam of amplitude E_{in} is sent towards the BS. The amplitude of the field transmitted by the beam splitter and then reflected by the NE after a round trip (RT), is

$$E_N^{RT} = -t_{BS} r_{NE} e^{-i2kL_N} E_{in} \tag{2.2}$$

where $k = \frac{2\pi}{\lambda}$ is the wave number of the light with λ light wavelength, t_{BS} is the transmittivity of the BS and r_{NE} is the reflectivity of the North End mirror. The amplitude of the field reflected by the beam splitter and then reflected by the WE after a round trip (RT), is

$$E_W^{RT} = r_{BS} r_{WE} e^{-i2kL_W} E_{in} \tag{2.3}$$

where r_{BS} is the reflectivity of the beam splitter, r_{WE} is the reflectivity of the West End mirror. These two fields recombine into the 50/50 beam splitter. The resulting field at the output port of the detector has an amplitude of

$$E_{out} = -r_{BS}E_N^{RT} + t_{BS}E_W^{RT} = r_{BS}t_{BS}\left(r_{NE}e^{-i2kL_N} + r_{WE}e^{-i2kL_W}\right)E_{in}$$
$$= r_{BS}t_{BS}E_{in}e^{-ik(L_N+L_W)}\left[r_{NE}e^{-ik(L_N-L_W)} + r_{WE}e^{ik(L_N-L_W)}\right]. \tag{2.4}$$

The reflectance and the transmittance are defined as $|r_i|^2 = R_i$ and $|t_i|^2 = T_i$ respectively. The output power detected by the photodetector is

$$P_{out} = |E_{out}|^2 = P_{in}R_{BS}T_{BS}(R_{NE} + R_{WE} + 2r_{NE}r_{WE}\cos\phi) \tag{2.5}$$

where

$$\phi = 2k(L_N - L_W) = 2k\delta L. \tag{2.6}$$

The output power is proportional to the differential length of the two arms.
The maximum detected power corresponds to $\phi = 0$

$$P_{max} = P_{in}R_{BS}T_{BS}(R_{NE} + R_{WE} + 2r_{NE}r_{WE}) \tag{2.7}$$

and the minimum power to $\phi = \pi$

$$P_{min} = P_{in}R_{BS}T_{BS}(R_{NE} + R_{WE} - 2r_{NE}r_{WE}). \tag{2.8}$$

The *contrast* can be introduced from Eqs. 2.7 and 2.8

$$C = \frac{P_{max} - P_{min}}{P_{max} + P_{min}} = \frac{r_{NE}r_{WE}}{R_{NE} + R_{WE}}. \tag{2.9}$$

The output power as a function of C is

$$P_{out} = P_{in}R_{BS}T_{BS}(R_{NE} + R_{WE})(1 + C\cos\phi). \tag{2.10}$$

P_{out} depends only by the phase ϕ, that can also be expressed as

$$\phi = \phi_0 + \phi_{GW}(t) \tag{2.11}$$

where ϕ_0 is the static term and $\phi_{GW}(t)$ is the time—dependent phase introduced by the passing gravitational wave. If $R_{NE} = R_{WE}$ and $\phi_0 >> \phi_{GW(t)}$, Eq. 2.10 can be re-written as

$$P_{out} = \frac{P_{in}}{2}\left(1 + C\cos\left(\phi_0 + \phi_{GW}(t)\right)\right) = \frac{P_{in}}{2}\left(1 + C\cos\phi_0 - \phi_{GW}(t)\sin\phi_0\right). \tag{2.12}$$

This corresponds to a change in power introduced by the passage of the gravitational wave equal to

$$\delta P_{GW}(t) = \frac{P_{in}}{2} \cdot C \cdot \phi_{GW}(t) \cdot \sin\phi_0. \tag{2.13}$$

If the static part is zero, starting from the definition in Eq. 2.6, the passage of a gravitational wave introduces a relative deformation of the arms $\delta L_{GW}(t)$ which is revealed as a difference in phase

$$\phi_{GW}(t) = 2k\delta L_{GW}(t) = \frac{4\pi}{\lambda}\delta L_{GW}(t). \tag{2.14}$$

Equation 2.13 can be re-written using Eqs. 2.14 and 2.1, as

$$\delta P_{GW}(t) = \frac{P_{in}}{2} \cdot C \cdot \sin\phi_0 \cdot \frac{2\pi}{\lambda} \cdot h(t) \cdot L \tag{2.15}$$

where $L = L_N = L_W$.

Any variation caused by an alteration in the distance between the mirrors, produces a very small shift in phase between the beams and, thus, a variation of the intensity of the light, which is proportional to the wave's amplitude. The variation of light detected $\delta P_{GW}(t)$ is proportional to the length of the arms and to the input power P_{in}, meaning that to detect more signal it is needed to have long arms and an high input power. Two different configurations of the detector can be defined from the value of the static phase ϕ_0 in Eq. 2.12:

- $\phi_0 = \pi$: dark fringe, i.e. no power at the output port of the detector;
- $\phi_0 = 0$: bright fringe.

2.1.2 *Fabry-Perot Cavities*

In order to increase the effect that the passage of a gravitational wave has on the signal at the output port of the Michelson interferometer, a more complex configuration is used. Fabry-Perot cavities has been introduced along the interferometer arms in order to increase their effective length. The scheme of a simple Fabry-Perot cavity is shown in Fig. 2.3. The cavity is composed by two mirrors M_1 and M_2 of reflectivity and transmissivity $r_{1,2}$ and $t_{1,2}$.

The electromagnetic fields measured at different points of the cavity are

- E_{in} : the input field;
- $E_1 = t_1 E_{in} - r_1 E_4$: the input field transmitted by M_1;
- $E_2 = E_1 e^{-ikL}$: the field arriving on M_2;
- $E_3 = -r_2 E_2 = -r_2 E_1 e^{-ikL}$: the field reflected by M_2;

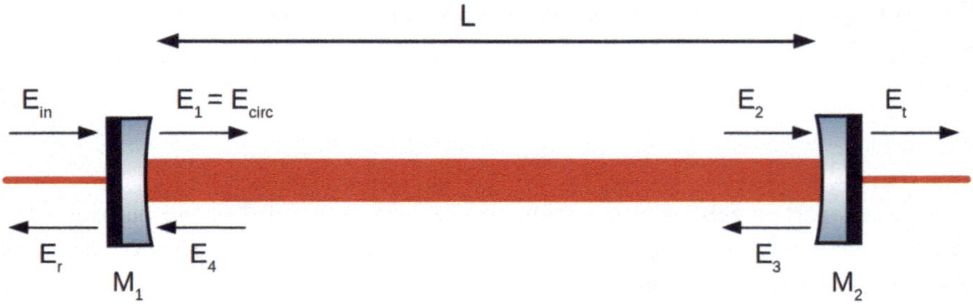

Fig. 2.3 Simple scheme of a Fabry-Perot cavity

- $E_4 = -r_2 E_2 e^{-ikL} = -r_2 E_1 e^{-i2kL}$: the field reflected by M_2 arriving back on M_1;
- $E_r = -r_1 E_{in} + t_1 E_4$: the field reflected by the Fabry-Perot cavity;
- $E_t = t_2 E_2 = t_2 E_1 e^{-ikL}$: the field transmitted by the Fabry-Perot cavity.

Combining these relations, the reflected, the transmitted and the circulating fields are derived:

- Field reflected by the cavity:

$$E_r = \frac{r_1 - r_2 e^{-i2kL}}{1 - r_1 r_2 e^{-i2kL}} E_{in} = r_{cav} E_{in} \tag{2.16}$$

  where r_{cav} is the reflectivity of the Fabry-Perot cavity;
- Field transmitted by the cavity:

$$E_t = \frac{t_1 t_2 e^{-ikL}}{1 - r_1 r_2 e^{-i2kL}} E_{in} = t_{cav} E_{in} \tag{2.17}$$

  where t_{cav} is the transmissivity of the Fabry-Perot cavity;
- Field circulating inside the cavity:

$$E_1 = E_{circ} = \frac{t_1}{1 - r_1 r_2 e^{-i2kL}} E_{in} = g_{cav} E_{in} \tag{2.18}$$

  where g_{cav} is the gain of the Fabry-Perot cavity.

The respective powers of these fields are shown in Fig. 2.4 for a 3 km long Fabry-Perot cavity.

The main quantities describing a Fabry-Perot cavity are [26]:

- Optical gain: amplification factor within the cavity

$$G = g_{cav}^2 = \left(\frac{t_1}{1 - r_1 r_2} \right)^2 \tag{2.19}$$

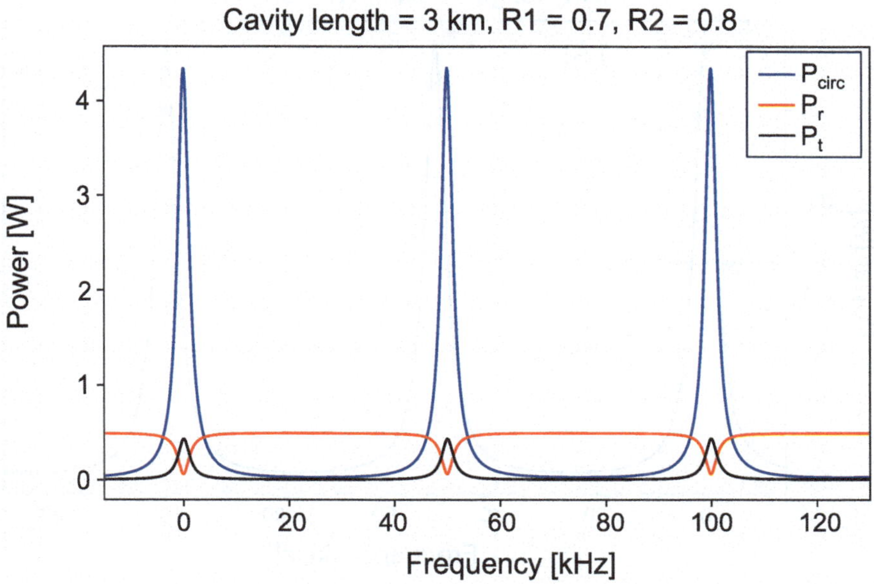

Fig. 2.4 Circulating (blue), transmitted (black) and reflected (red) powers by a 3 km long Fabry-Perot cavity with $R_1 = 0.7$ and $R_2 = 0.8$ as a function of the frequency

- Finesse: it is a measure of how wide the resonance peaks are in relation to their frequency distance and it can be expressed as a function of the mirror reflectivities when $r_1, r_2 \sim 1$

$$\mathcal{F} = \frac{\pi\sqrt{r_1 r_2}}{1 - r_1 r_2} \tag{2.20}$$

- Storage time: it is the average time spent by a photon inside the cavity before exiting through one of the mirrors

$$\tau = \mathcal{F}\frac{2L}{\pi c} = N\frac{L}{c} = \frac{\pi}{2 f_p} \tag{2.21}$$

  where N is the average number of round trip inside a Fabry-Perot cavity and f_p is the Fabry-Perot cavity pole cut frequency.
- Frequency response of the Fabry-Perot cavity:

$$g(f) = \frac{1}{1 + \left(\frac{f}{f_p}\right)^2} \tag{2.22}$$

- Free spectral range (FSR): it is the frequency distance between two successive resonances (see Fig. 2.5)

$$FSR = \Delta\nu = \frac{c}{2L} \tag{2.23}$$

  where c is the speed of light;

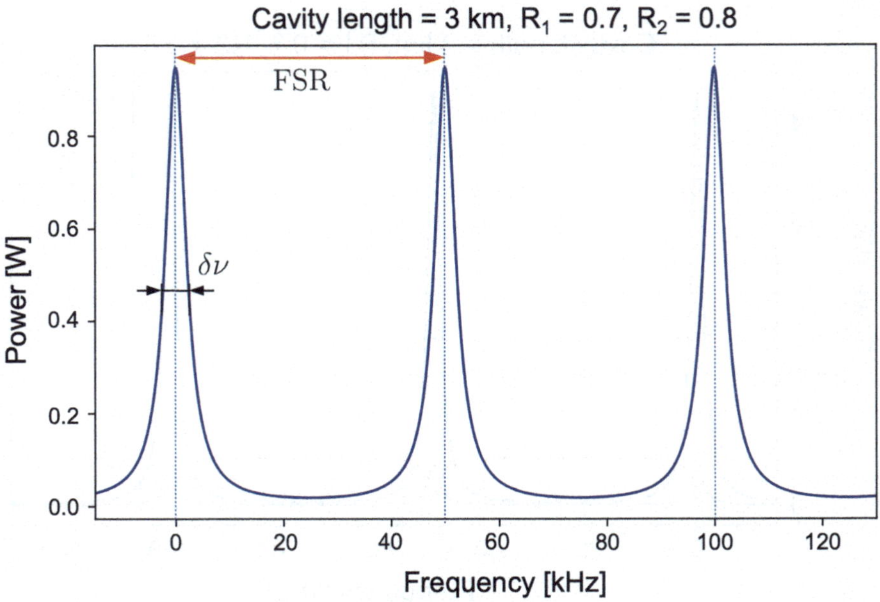

Fig. 2.5 Power transmitted by a 3 km long Fabry-Perot cavity with $R_1 = 0.7$ and $R_2 = 0.8$ as a function of the frequency. The Free Spectral Range (FSR) is defined as the distance between two consecutive maxima. The linewidth $\delta\nu$ is full width half-maximum (FWHM) of the resonance peak

- Linewidth: it is the full width half-maximum (FWHM) of the resonance peak (see Fig. 2.5)

$$\delta\nu = \frac{\Delta\nu}{\mathcal{F}}. \tag{2.24}$$

The power circulating inside the Fabry-Perot cavity, from Eq. 2.18, is given by

$$P_{circ} = |E_{circ}|^2 = \frac{t_1^2}{1 - r_1 r_2 + 4\sin^2(kL)} \cdot P_{in} = \frac{G}{1 + \left(\frac{2\mathcal{F}}{\pi}\right)^2 \sin^2(kL)} \cdot P_{in} \tag{2.25}$$

where $kL = \frac{\nu}{c}L$, introducing the dependency on the frequency.

The resonance condition of the cavity occurs when the phase shift of the field after a round trip corresponds to an integer number of π, i.e. $\phi = kL = n\pi$.

The two Fabry-Perot cavities along the interferometer arms are shown in Fig. 2.6. The phase detected at the output port, using Eq. 2.16 instead of Eq. 2.2 in Eq. 2.4, can be written as

$$\phi = \phi_0 + \phi_{GW}(t) = \phi_0 + \frac{4\pi}{\lambda}\frac{2\mathcal{F}}{\pi}\delta L_{GW}(t) = \phi_0 + N\frac{4\pi}{\lambda}\delta L_{GW}(t) \tag{2.26}$$

where ϕ_0 represents the phase at the resonance.

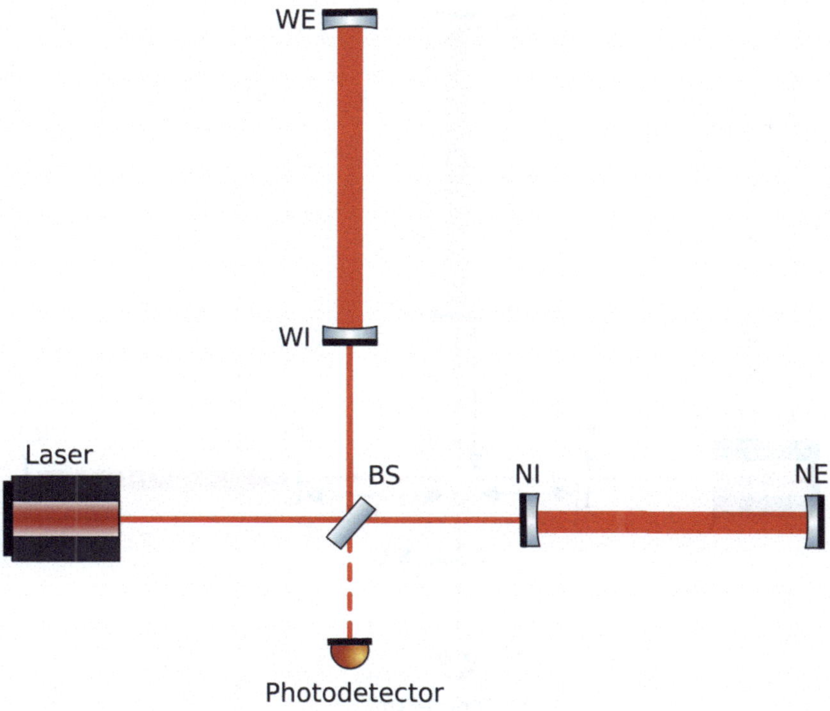

Fig. 2.6 Michelson interferometer with Fabry-Perot cavities along the arms. The Fabry-Perot cavity on the West arm is composed by the West Input mirror (WI) and the West End mirror (WE). The one on the North arm is composed by the North Input mirror (NI) and the North End mirror (NE)

This means that the small shift of the mirrors introduced by the passage of a gravitational wave is amplified by the Fabry-Perot cavity. The amplification factor is proportional to the Finesse of the interferometer arms.

2.1.3 Power Recycling Cavity

Since the output power variation introduced by the gravitational wave is proportional to the input power (see Eq. 2.13), increasing the power circulating in the interferometer improves its performance. Furthermore, the detector operates close to the dark fringe condition ($\phi_0 << \pi$ in Eq. 2.13), i.e. there is almost no light coming out to the output port and this means that all the power is sent back to the laser. The power recycling technique can therefore be used by inserting a mirror (PR) between the laser and the Beam Splitter (BS) to send the light back to the interferometer, as shown in Fig. 2.7. The effect is that the interferometer input beam sees a Fabry-Perot cavity formed by the PR as an input mirror and the rest of the interferometer as an

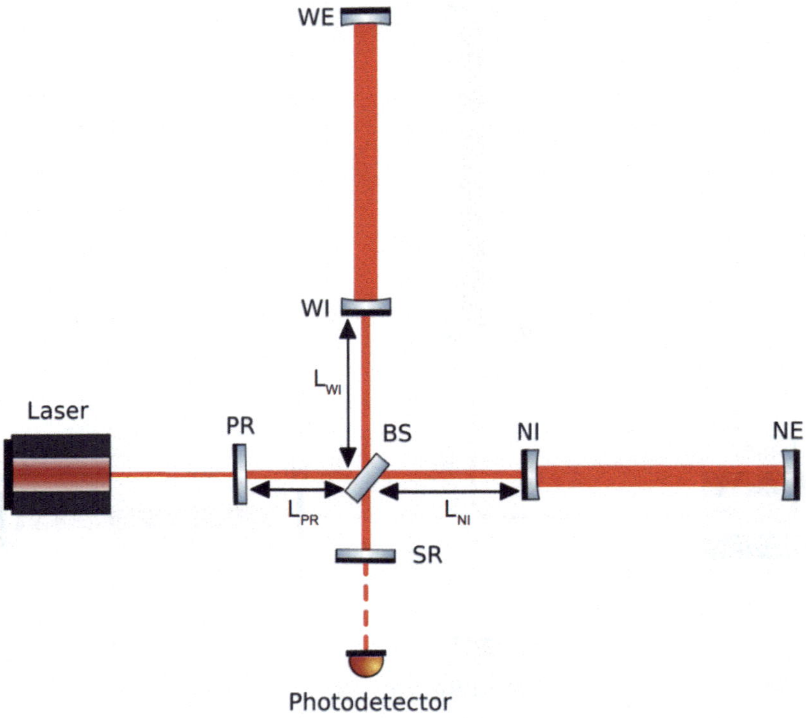

Fig. 2.7 Michelson interferometer with Fabry-Perot cavities along the arms and two additional mirrors: the Power Recycling (PR) and the Signal Recycling (SR). The PR amplifies the light circulating in the interferometer. L_{PR} is the distance between the PR mirror and the Beam Splitter (BS). L_{NI} and L_{WI} are the distances between the North Input and the West input mirrors fro the BS. The SR amplifies the signal arriving at the detection port

output mirror. Defined L_{PR} as the distance between the PR and the BS, the field circulating inside the cavity is given by (from Eq. 2.18)

$$E_{PR} = \frac{t_{PR}}{1 - r_{PR} r_+ e^{-i2kL_{PR}}} E_{in} \tag{2.27}$$

where r_+ is given by

$$r_+ = -r_{FP} e^{-ik(L_{NI}+L_{WI})} \tag{2.28}$$

and r_{FP} is the Fabry-Perot reflectivity. The circulating power can be obtained as in Eq. 2.25 and the resulting optical gain is (using Eq. 2.19)

$$G_{PR} = \left(\frac{t_{PR}}{1 - r_{PR} r_+} \right)^2 . \tag{2.29}$$

It represents the power amplification introduced in the interferometer by adding PR mirror.

2.1.4 Signal Recycling Cavity

The final detector configuration is a dual-recycled Michelson interferometer, where in addition to the power recycling there is another recirculation cavity, the signal recycling (SRC) one. This cavity can be described by considering the entire power-recycled interferometer as a single optical system and the signal recycling mirror (SR) as the second element of this cavity, positioned at the output port of the detector. The reflectivity and the transmissivity of the SR are denoted by r_{SR} and t_{SR} and the distance between the signal recycling mirror and the power recycled interferometer is denoted by L_{SR}.

The signal recycling technique is generally adopted to increase the interferometer sensitivity or band-width to differential displacement. Only when a differential motion is present, a sideband signal at a certain frequency f is transmitted at the detector's antisymmetric port and its resonance condition in the SRC is altered. The sideband signal, which is generated in the long arms of the interferometer, sees an effective input mirror composed of the cavity made by SR and the Fabry-Perot input mirror. Thus tuning the SRC can change the reflectivity of the input mirror that the sideband signal sees and thus change the position of the Fabry-Perot cavity pole, defined in Eq. 2.1.2.

The field transmitted by the signal recycling mirror at the anti-symmetric port is given by Vajente [27]

$$E_{t_{SR}}(f) = \frac{t_{PR} \sin\left(\frac{f \Delta L}{c}\right) e^{-i\phi \pm i \frac{f}{c} L_{SRC}}}{1 - r_{PR} \cos\left(\frac{f \Delta L}{c}\right) - r_{SR}\left(\cos\left(\frac{f \Delta L}{c}\right) - r_{PR}\right) e^{-i2\phi \pm i2\frac{f}{c} L_{SRC}}} E_{in} \tag{2.30}$$

where the following quantity are defined as

- signal recycling cavity length

$$L_{SRC} = L_{SR} + \frac{L_{NI} + L_{WI}}{2} \tag{2.31}$$

- Schnupp asymmetry: static length difference between the small interferometer arms which is, in the case of Virgo, given by

$$\Delta L = L_{NI} - L_{WI} = 23\text{cm} \tag{2.32}$$

- phase tuning of the signal recycling cavity

$$\phi = kL_{SRC} + \frac{\pi}{4} \tag{2.33}$$

Depending on the value of ϕ, the interferometer is in different configurations:

- $\phi \in]-\pi, 0[$ gives a resonant enhancement of the sensitivity at a single frequency (*detuned signal recycling*);
- $\phi = \pm\pi$ gives an enhancement of the lower frequency gain (*tuned signal recycling*);
- $\phi = 0$ gives a broad-band response of the interferometer (*resonant sideband extraction* regime), which is the configuration used in Virgo;
- $\phi \in]0, \pi[$ gives a low frequency gain increase.

2.2 Noise Sources

Ground-based gravitational wave detectors must be sensitive to extremely small differential variations of their arms length, 1000 times smaller than a proton. Various noise sources mask these signals, mimicking the gravitational wave signal. It is therefore necessary to understand and reduce these noises as much as possible.

The noise can be described as a random time series *n(t)* and its single-sided Power Spectral Density (PSD) is given by

$$S(f) = \lim_{T\to\infty} \frac{2}{T} \left| \int_{-T}^{+T} n(t) e^{-2i\pi f t} dt \right|^2 . \tag{2.34}$$

This quantity describes the variation of the time serie *n(t)* at the frequency *f*. The Amplitude Spectral Density (ASD) is given by

$$h(f) = \sqrt{S(f)}. \tag{2.35}$$

The PSD associated to the noise n will be indicated as S_n while its ASD as h_n.

The noises can be divided in two group: fundamental and technical noises.

The fundamental noises are the ones intrinsic to the interferometer design. The main noises involved are the seismic, the thermal and the quantum noises (see Sects. 2.2.1, 2.2.3, 2.2.4, and 2.2.5). The spectra of these noises are shown in Fig. 2.8.

The technical noises arise from all the techniques and related equipment put in place for the interferometer control and the signal detection. In this category there are scattered light noise, electronic noise, etc (see Sect. 2.2.6).

2.2.1 Seismic Noise

Seismic noise is due to ground vibrations coupling with the test masses. To attenuate this noise, the interferometer mirrors are decoupled from the ground by a complex suspension system, see Fig. 2.9. This noise comes from several sources. For example

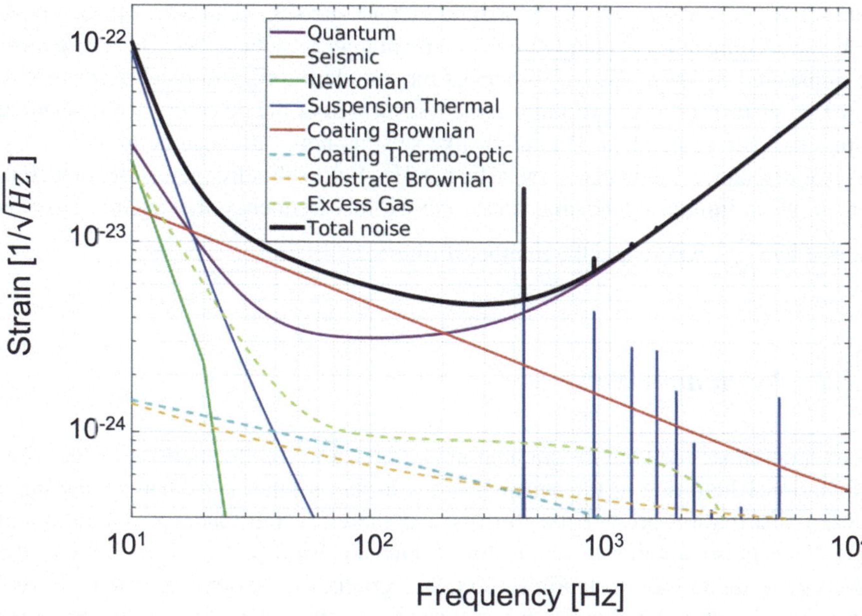

Fig. 2.8 Sensitivity curve of Advanced Virgo Plus with 40 W of injected power. The fundamental noises are underlined while the technical noises are responsible for the total noise curve (in black) to be higher than the estimated noises

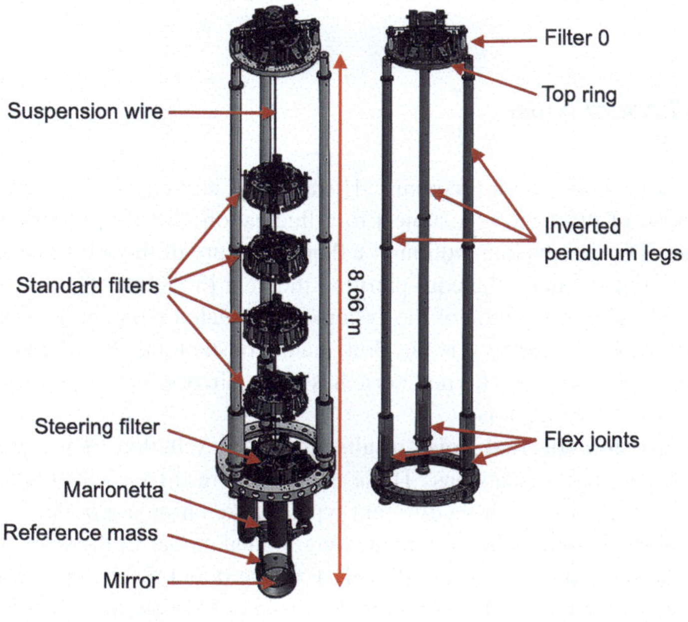

Fig. 2.9 Virgo super attenuator [30]

below 1 Hz, it is mainly caused by natural factors, such as ocean waves [28]. Above 1 Hz the seismic noise is mainly produced by human activities [29]. This noise must be attenuated by more than 10 orders of magnitude in order to reach a sensitivity of 10^{-23} at 100 Hz. For this purpose the test masses of the detectors are suspended with a cascade of pendula. Pendulum-like seismic noise attenuators are conceived to have resonance frequencies lower than 1 Hz. Below this frequency the detectors sensitivity is limited by seismic noise. Above this frequency the seismic noise is filtered by $\frac{1}{f^{2n}}$, where n is the number if filters.

2.2.2 Newtonian Noise

Newtonian noise (or Gravity gradient noise) is due to the fluctuations of the local gravitational field due to the seismic waves in the ground, that changes the mass density distribution around the mirrors and therefore introduces a displacement [31]. This noise dominates in the low frequency band and acts directly on the mirrors bypassing their isolation system, so a ground-based detector can not have a frequency detection band below this frequency. A way to beat this limit is to locate the instrument in a quiet place, for example underground as it has been done for the KAGRA detector [32]. Indeed, it has been observed that the effects of the newtonian noise are reduced by about two orders of magnitude by going underground. In addition, one method of reducing Newtonian noise in current gravitational wave detectors is to measure it and subtract it from the detector's output data [33].

2.2.3 Thermal Noise

The Fluctuation-Dissipation theorem [34] states that any physical system affected by some form of dissipation is subject to a fluctuation directly proportional to its temperature. This results in a motion of a thermal nature of the mirrors and of their suspensions. Furthermore, the equi-partition theorem [34] states that every normal oscillation and vibration mode of the system is associated with energy fluctuations proportional to $k_B T$ where k_B is the Boltzmann constant and T is the equilibrium temperature of the system. Thermal noise has two main origins: suspension thermal noise and mirror thermal noise.

The **suspension thermal noise** results from the excitation of the mechanical modes of the mirror suspensions. These excitations result in a fluctuation of the position of the test masses. Two different types of excitation are possible. The first is the pendulum thermal noise, where the longitudinal modes of the suspensions are excited at the resonance frequency (0.6 Hz for Virgo) and results in a longitudinal displacement of the mirror. This noise is dominant at frequencies below 50 Hz and has been reduced in the past by replacing the steel suspension wires with monolithic fused silica wires. The second consists of the excitation of the vibrational modes of the suspensions wires, which resonant frequencies are ∼450 Hz and its harmonics.

The **mirror thermal noise** can be due to both the vibration modes of the substrates of the mirrors and of their coating, which is the major contributor. Coating thermal noise is typically dominant between 30 and 100 Hz.

One effective way to reduce this noise is to put the detector mirrors in a cryogenic environment, as was done with KAGRA [35] and is planned for future detectors.

2.2.4 *Shot Noise*

The gravitational wave signal is extracted from the measurement of power reaching the antisymmetric port of the detector. The statistical fluctuation in the power is coupled directly to the system, introducing what is called *shot noise*. More specifically, this noise arises from the quantum nature of the light being detected. In this section the semi-classical approach is followed. The photons arriving on the photo-detector, denoted by N, are discrete and independent events and they follow a Poissonian distribution

$$P(N) = \frac{\overline{N}^N}{N!} e^{-\overline{N}}. \tag{2.36}$$

where $\overline{N}$ is the average number of photons and $\sigma_N^2 = \overline{N}$ is the variance, considering that $\overline{N} >> 1$. Considering the laser energy in terms of photons as $E_{ph} = \hbar\omega$, with $\hbar$ being the reduced Plank constant, the rate of measured photons $\overline{n}$ can be defined as

$$\overline{n} = \frac{P_{out}}{E_{ph}} = \frac{P_{out}}{\hbar\omega}, \tag{2.37}$$

the average number of measured photons over a time T as

$$\overline{N} = \sigma_N^2 = \overline{n}T = \frac{P_{out}}{\hbar\omega} T \tag{2.38}$$

and the statistical fluctuations as

$$\sigma_N = \sqrt{\overline{N}}. \tag{2.39}$$

Therefore, from Eq. 2.12, the power fluctuation in the time interval T can be characterized as

$$\delta P_{shot} = \frac{\hbar\omega}{T}\sigma_N = \hbar\omega\sqrt{\frac{P_{out}}{\hbar\omega T}} = \sqrt{\frac{P_{in}}{2T}(1 + C\cos\phi)\,\hbar\omega}. \tag{2.40}$$

The signal-to-noise ratio in the frequency domain, for a Michelson interferometer, can be defined from the Eqs. 2.40 and 2.13 as

$$SNR(f) = \frac{\delta P_{GW}(f)}{\delta P_{shot}(f)} = \frac{1}{2}\sqrt{\frac{P_{in}}{\hbar\omega}}\frac{C\sin\phi_0}{\sqrt{1+C\cos\phi_0}}\phi_{GW}(f). \tag{2.41}$$

which is maximum for

$$\cos\phi_0 = \frac{-1+\sqrt{1-C^2}}{C}. \tag{2.42}$$

Therefore the best SNR is found at the dark fringe condition assuming C ~ 1 $(\cos\phi_0 = -1)$.

From Eqs. 2.1 and 2.26, $\phi_{GW}(f)$ can be expressed as

$$\phi_{GW}(f) = \frac{2\pi L}{\lambda}\frac{2\mathcal{F}}{\pi}h(f). \tag{2.43}$$

The strain corresponding to the minimum detectable gravitational wave amplitude for a power recycled Fabry-Perot Michelson interferometer limited by shot noise (i.e. SNR = 1) is therefore

$$h_{shot}(f) = \frac{\lambda}{4\mathcal{F}L}\sqrt{\frac{\hbar\omega}{P_{in}G_{PR}}}\frac{1}{\sqrt{g(f)}} \tag{2.44}$$

where $g(f)$ is defined in Eq. 2.22.

2.2.5 Radiation Pressure Noise

The radiation pressure noise is due to the momentum transferred onto the test masses by the unequally spaced in time photons arriving on them. The transferred momentum introduces a fluctuating radiation pressure force on the suspended mirrors that is related to fluctuations in power of the incident beam according to

$$\delta F = \frac{2\delta P_{in}}{c} \tag{2.45}$$

where it has been assumed that the power arriving on the mirror is $P_{in}/2$. Suspended mirrors of mass M are treated as harmonic oscillators far from their resonances, so that the displacement induced on the mirror at angular frequency f is

$$\delta x(f) = \frac{\delta F(f)}{M(2\pi f)^2} = \frac{1}{Mf^2}\sqrt{\frac{\hbar P_{in}}{4\pi^2 c\lambda}}, \tag{2.46}$$

where a simple Michelson configuration is considered. The fluctuations in the two arms are anti-correlated, i.e. an extra photon in one arm corresponds to a missing photon in the other. Thus the induced variation of the average arm length L is $2x(f)$. The radiation pressure noise for a power recycled Fabry-Perot interferometer is given by

$$h_{rp}(f) = \frac{2}{L}x(f) = \frac{4\mathcal{F}}{ML}\sqrt{\frac{\hbar P_{in} G_{PR}}{\pi^5 c\lambda}}\frac{\sqrt{g(f)}}{f^2}. \tag{2.47}$$

Total quantum noise strain is given by the sum in quadrature of shot noise and radiation pressure noise

$$h_{qn} = \sqrt{h_{shot}^2 + h_{rp}^2}. \tag{2.48}$$

The radiation pressure noise follows $1/f^2$ and thus predominates at low frequencies, while the shot noise has a pattern that does not depend directly on the frequency, but only on the frequency of the Fabry-Perot cavity pole, see Eq. 2.44, and thus dominates at high frequencies. In addition, the radiation pressure noise is proportional to $\sqrt{P_{in}}$, while the shot noise to $1/\sqrt{P_{in}}$. This shows that according to the semi-classical treatment, it is not possible to reduce both of these noises at the same time. For each frequency, one can find the power value that minimizes quantum noise, i.e. h_{qn}. This curve is called the *standard quantum limit* (SQL) and is defined as

$$h_{SQL}(f) = \frac{1}{f}\sqrt{\frac{\hbar}{M\pi^2 L^2}}. \tag{2.49}$$

The quantum noise overall spectrum is shown in Fig. 2.10.

However, a technique has been developed in order to go below the SQL, by introducing in the detector particular quantum states of light called *squeezed states*. This technique will be described in detail in Part II as it is the main subject of this thesis.

2.2.6 Technical Noises

In addition to the fundamental noises just described, which are intrinsic in the components of the interferometer, also the so-called *technical noises* are spoiling the detector sensitivity. Their impact can be reduced by a proper understanding of the coupling mechanism and an adequate technical design. With experience, several noises have been identified and efforts have being made to mitigate them as best as possible.

Currently recognized technical noises include: scattered light noise, electronic noise, laser noises (such as frequency noise, power noise etc), contrast defect, control noise.

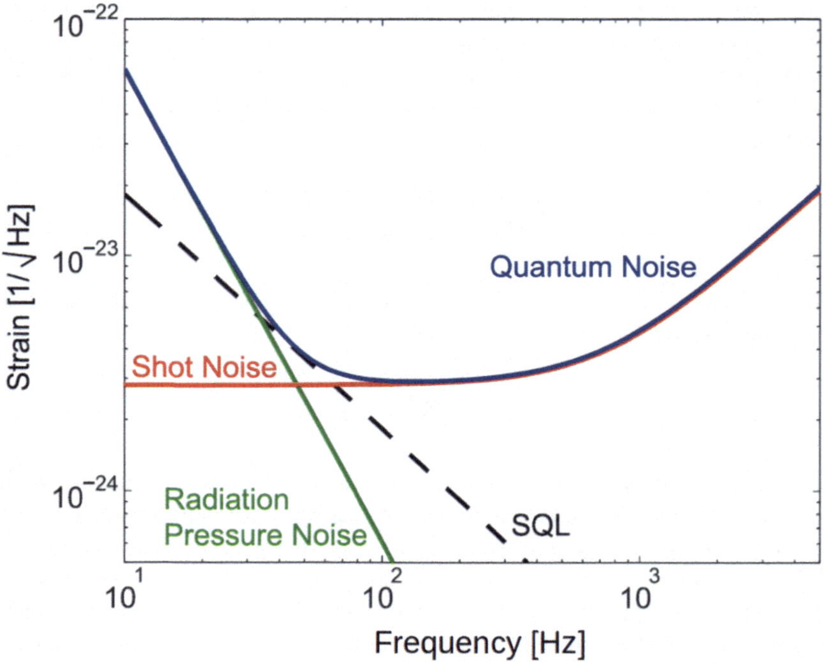

Fig. 2.10 Quantum noise spectrum: radiation pressure noise (green curve) dominating at low frequency and shot noise (red curve) dominating at high frequency. The overall effect of quantum noise on the sensitivity curve is expressed by the blue line. The standard quantum limit (SQL) is the dashed black curve. Figure from [38]

Scattered light noise, for example, is due to the fact that scattered light, as it re-couples with the detector's main beam, introduces low-frequency noise. Scattered light from the detector main mirrors was predicted and then mitigated by putting baffles along the arms [36], so that it could be stopped before it could re-couple with Virgo's main beam mode. It is known that also the optics on the suspended benches, used for interferometer control and gravitational wave detection, generate scattered light, producing a clearly visible effect on the sensitivity curve [37]. This effect will be described in Part III.

Moreover, in the shot noise calculation it has been assumed that the contrast is equal to 1, but it is actually slightly less. This is due to the fact that the two beams recombining on the beam splitter, do not have the same geometric properties due to the defects in the mirrors. This means that higher order modes which do not contain the gravitational wave signal arrive at the output port of the detector. This effect is called *contrast defect*. To make sure that only the carrier signal in its fundamental mode arrives at the photodiode, it is necessary to introduce a cavity that cleans the beam, the Output Mode Cleaner, as will be described in Part IV.

2.3 The Advanced Virgo Plus Detector

The European gravitational wave detector Virgo is located in the Tuscan countryside near Cascina, Italy. It takes its name from the Virgo cluster containing about 1500 galaxies, about 16.5 million Mpc far from the Earth, which is located in the Virgo constellation. The proposition to build this detector was made by the scientists Adalberto Giazotto and Alain Brillet in 1987. Its construction began ten years later and it was inaugurated in 2003. Virgo is a power recycled Michelson interferometer with Fabry-Perot cavities along its 3 km long arms [39], as described in Sect. 2.1.3. Several upgrades were needed to improve its sensitivity and thus to detect gravitational waves. The first upgrade Advanced Virgo lead to the first three-detectors detection in the August 2017 [40]. Advanced Virgo remained in operation until March 2020. From then until spring 2023, the interferometer is in the commissioning phase. This upgrade is called Advanced Virgo Plus (AdV+).

2.3.1 Upgrades Towards Advanced Virgo Plus

The main goal of AdV+ is to reduce the detector's sensing noise, noise that comes from quantum noise (see Sects. 2.2.4 and 2.2.5) by introducing three main changes to the Advanced Virgo configuration:

- the installation of the signal recycling mirror that allows the detection bandwidth to be increased by reshaping the quantum noise (see Sect. 2.1.4);
- the increase of the laser input power to reduce the effect of shot noise;
- the injection of frequency-dependent squeezing to further reduce shot noise without increasing radiation pressure noise at low frequencies (see Sects. 2.2.4 and 2.2.5).

The optical configuration of AdV+ is shown in Fig. 2.11. The main parameters of AdV+, compared with the ones of AdV are reported in Table 2.1.

AdV+ is a double-recycled Michelson interferometer, with Fabry Perot cavities along the 3 km long arms, as described in Sect. 2.1. On the optical layout are indicated the photodiodes detecting the interferometer beams, whose signals are used to extract the GW signal and to control the interferometer's longitudinal and angular degrees of freedom: B1, B1p, B2, B4, B5, B7 and B8. Each sensor is located on a suspended bench, which is not shown in the figure for simplicity but are listed in Table 2.2.

The laser input beam passes through a 143-m-long triangular input mode cleaner cavity and then through a Faraday Isolator which prevents light reflected by the interferometer from re-entering the laser. The beam is split by the 50/50 beam splitter before entering the Fabry-Perot cavities along the two arms of the

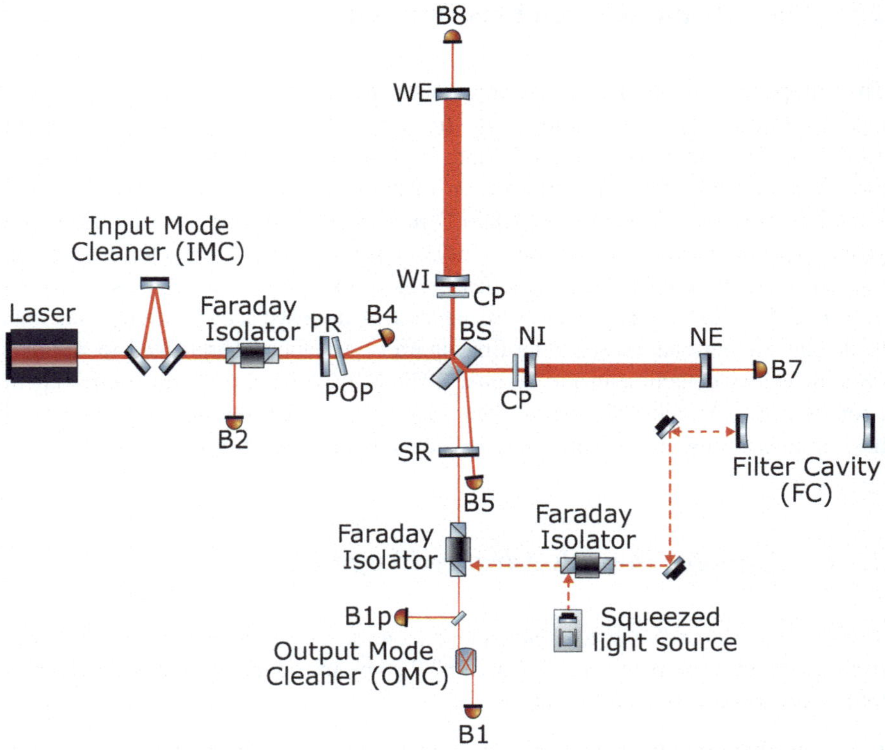

Fig. 2.11 Advanced Virgo Plus optical configuration

interferometer.[2] The beams reflected by the North and West Fabry-Perot cavities recombine on the beam splitter, close to the dark fringe condition (see Sect. 2.1). This mean that the beam at the output port has a low power ($\sim$ 1 W). It passes through the signal recycling mirror (SRM) and propagates through the output Faraday Isolator (on the detection bench SDB1). The main beam passes through the Output Mode Cleaner (OMC) cavity. The OMC is a small monolithic bow-tie Fabry-Perot cavity, whose aim is to filter the dark fringe beam removing the higher order modes and control sidebands (see Sect. 2.4.2) and to keep only the GW signal. More details on the OMC are given in Chap. 10. The beam transmitted by this cavity is called B1. A small pick-off, called B1p, bypasses the OMC, reaching directly the sensors. There are several other pick-offs used to monitor the powers in the ITF cavities and to control their length and alignment. The beam reflected by the Power Recycling Cavity (PRC) is called B2. The beam sensing the power inside the PRC

[2] In front of the input mirrors of the Fabry-Perot cavities there are the compensation plates (CPs), used to compensate for the thermal gradient introduced by the absorption of the beam in the substrates.

Table 2.1 Main parameters of AdV+ compared with the ones od AdV duing O3

Parameter	Advanced virgo (O3)	Advanced virgo plus (O4)
Laser and injection		
Laser power	100 W	130 W
Input power in the ITF	18 W	40 W
Input mode cleaner length	143 m	143 m
Input mode cleaner finesse	1000	1000
ITF optical configuration		
Arm cavity length	3 km	3 km
Arm cavity finesse	450	450
Power recycling gain	39	39
Arm circulating power	201 kW	447 kW
Signal recycling	None	Yes
Mirrors		
Beam splitter	55 cm × 6.5 cm, 34 kg, T=50%	55 cm × 6.5 cm, 34 kg, T=50%
Input test masses	35 cm × 20 cm, 42 kg, T=1.4%	35 cm × 20 cm, 42 kg, T=1.4%
End test masses	35 cm × 20 cm, 42 kg, T=5 ppm	35 cm × 20 cm, 42 kg, T=5 ppm
Power recycling mirror	35 cm × 10 cm, 21 kg, T=5%	35 cm × 10 cm, 21 kg, T=5%
Signal recycling mirror	None	35 cm × 10 cm, 21 kg, T=40%
Detection		
Output mode cleaner	Double cavity, finesse = 123	Single cavity, finesse = 1000
Detection losses	15%	< 10%
Quantum noise reduction system		
Squeezed vacuum source	8 dB	12 dB
Filter cavity	None	L = 285 m, finesse = 10000
Injection losses	11%	< 10%
Phase noise	120 mrad	40 mrad

Table 2.2 List of all the main Virgo beams, indicating the bench where they are detected. (*) SIB2 has been suspended for the first time during the O4 commissioning

Beam	Bench hosting the sensors
B1	Suspended detection bench 2 (SDB2)
B1p	Suspended detection bench 2 (SDB2)
B2	Suspended injection bench 2* (SIB2)
B4	Suspended power recycling bench (SPRB)
B5	Suspended detection bench 2 (SDB2)
B7	Suspended north end bench (SNEB)
B8	Suspended west end bench (SWEB)

is called B4 and it is extracted by the Pick-Off Plate (POP). The POP is tilted by an angle of 6° with respect to the PRM. The beams transmitted by the Fabry-Perot arm cavities are B7 and B8 respectively for the North arm and West arm. The beam reflected by the North Arm cavity is called B5. The detection system represents also the interface between the squeezing system and the interferometer, as it will be detailed in Chap. 9. The squeezing is injected through the Faraday Isolator on SDB1 and then enters the detector through its output port.

2.4 Feedback Control Loop

The ITF has to be kept in a precise working position in order to correctly operate and thus enable the detection of gravitational waves. This means that the suspended mirrors and benches must be positioned in the configuration that allows the power amplification in the resonant cavities and to be close to the dark fringe configuration (see Sect. 2.1.1). However, seismic noise causes the suspended elements to move if they are not permanently controlled. Therefore, dedicated *feedback control loops* have to be implemented. The control theory of dynamic systems is fundamental to Virgo's operation, so we provide the main details in Sect. 2.4.1. In addition, the longitudinal control of a cavity will be analyzed using a widely used technique called Pound-Drever-Hall.

2.4.1 Feedback Control Loop Principle

A *feedback control loop* is the process that takes the output of a system and feeds it back to its input to keep the system always at a given working point.

Physical systems are characterized by their ability to accept a parameter as input and produce an output in response. The systems considered in the following are linear, i.e. the input and output are linked by a linear relationship, and time invariant, i.e. their response does not change as a function of time. This type of system is called a *linear time-invariant system (LTI)* and can be described by a differential equation of the type:

$$a_0 \cdot x(t) + a_1 \cdot \frac{dx(t)}{dt} + a_2 \cdot \frac{d^2x(t)}{dt^2} + \ldots + a_n \cdot \frac{d^nx(t)}{dt^n} = b_0 \cdot y(t) \tag{2.50}$$

where *y(t)* is the input of the system and *x(t)* is the output of the system.

However, high-order differential equations can be difficult to solve in the time domain. One way to solve them is to consider the frequency domain and perform the Laplace transform and then recall the time domain response. The Laplace transform in the s-domain is defined by

$$\mathcal{L}\{f(t)\} = F(s) = \int_0^{\infty} f(t) \cdot e^{-st} dt \tag{2.51}$$

where s is a complex number. The behavior of an LTI system in the s-domain is described by its Transfer Function (TF), which consists of the ratio between the system's input $Y(s)$ and output $X(s)$, in the s-domain:

$$G(s) = \frac{X(s)}{Y(s)}. \tag{2.52}$$

The TF of the system can be described by rational functions such as

$$G(s) = g \cdot \frac{(s - z_1) \cdot (s - z_2) \cdot \ldots \cdot (s - z_N)}{(s - p_1) \cdot (s - p_2) \cdot \ldots \cdot (s - p_N))} \tag{2.53}$$

where g is the gain, z_i are the values that null the numerator of the TF and are called *zeros* and p_i are the values that null the denominator and are called *poles*. This is the great advantage of solving differential equations in the s-domain.

Most TFs can be described as the product of simple or second-order poles and zeros as in Eq. 2.53. The form of poles and zeros at frequency f_0 at first order is given by

$$Z_1(s) = s - 2\pi f_0 \tag{2.54}$$

$$P_1(s) = \frac{1}{s - 2\pi f_0}. \tag{2.55}$$

A simple pole is called an *integrator* because at the output the signal is the integral of the input signal. A simple zero is called a *differentiator* because the output is the derivative of the input.

Poles and zeros of second order at frequency f_0 are described by

$$Z_2(s) = s^2 + \frac{2\pi f_0}{Q} \cdot s + (2\pi f_0)^2 \tag{2.56}$$

$$P_2(s) = \frac{1}{s^2 + \frac{2\pi f_0}{Q} \cdot s + (2\pi f_0)^2} \tag{2.57}$$

where Q is the quality factor noticing that

- $Q = 0$ gives a simple pole/zero;
- $Q > 0.5$ gives an overshoot at $f < f_0$ and for $f > f_0$ TF follows a f^2 (zeros) or f^{-2} (poles) dependance.

The graphical representation of the transfer function of a system described with a zero or a pole, with different Q factors is shown in Fig. 2.12.

A feedback control loop can be described with this formalism by introducing several elements:

- the system to be controlled called *plant*, P(s), which in our case can be the length of a cavity;
- the *sensor*, S(s), that is used to extract the error signal: in our case, mainly photodiodes, quadrant photodiodes[3] and cameras are used;

[3] *Quadrant photodiodes* are discrete sensors composed by four optically active areas separated by small gaps. They are used to detect the position of the laser beam arriving on their sensitive area. In this thesis they will be called simply *quadrants*.

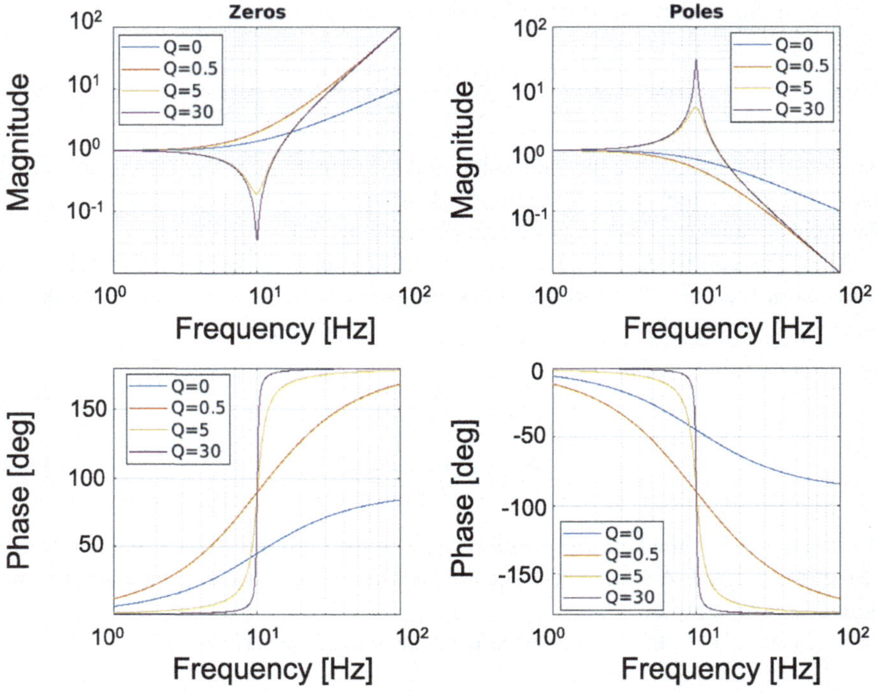

Fig. 2.12 On the left, the transfer function of a second order system described by a zero with different Q values, both the magnitude and the phase, is shown. On the right, the transfer function of a second order system described by a pole with different Q values, both the magnitude and the phase, is shown

- the *error signal*: the signal detected by the sensor which gives the information on the system's state. This signal must be bipolar, i.e. it must change sign around the point of interest, in order to give information on the direction of motion (a widespread technique is that of Pound-Drever-Hall to linearize the error signal, see Sect. 2.4.2);
- the feedback or *correction filter*, F(s): this is used to calculate the correction required to bring the system to the desired operating point;
- the *actuator*, Act(s): an element that acts on the plant to bring the system around the working point by applying the correction calculated by the filter F(s).

The schematic of a feedback control loop is shown in Fig. 2.13. The system's input is a perturbation, or noise, Y(s) that drives the plant away from its working point. The system output, X(s), is the error signal. The TF of the system measured when the loop is closed, i.e. when the feedback is sent to the plant, is called *Closed loop TF* and it is therefore given by

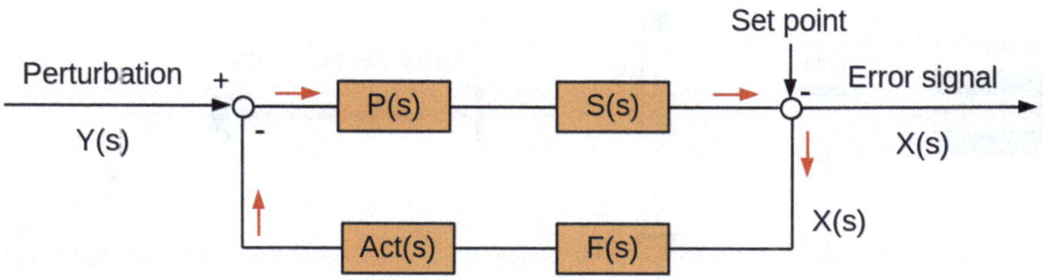

Fig. 2.13 Feedback control loop scheme

$$TF_{CL}(s) = \frac{X(s)}{Y(s)} = \frac{P(s) \cdot S(s)}{1 + P(s) \cdot S(s) \cdot F(s) \cdot Act(s)} = g_{CL}(s) \cdot P(s) \cdot S(s) \tag{2.58}$$

where the noise suppression exerted by the loop is quantified by the factor $g_{CL}(s)$, which is called *Closed loop gain*. The $g_{CL}(s)$ function gives information on the frequency range in which the loop is acting to reduce the input noise. The frequency for which this gain factor is 1 is called the *Unitary Gain Frequency (UGF)*.

Furthermore, for the feedback loop to function as desired, it must be stable. The system is stable if when an input disturbance is applied, its output converges. Stability is given by the behavior of the denominator of Eq. 2.58: if the product $P(s) \cdot S(s) \cdot F(s) \cdot Act(s)$ becomes equal to -1, the TF diverges and the system therefore becomes unstable. The product of the feedback loop elements is called *Open loop Transfer Function (OLTF)* and corresponds to the behavior of the system when the feedback is disconnected from the input.

More details on control theory can be found in [41].

2.4.2 Pound-Drever-Hall Technique: Longitudinal Sensing

An example of the system the be controlled in gravitational wave detectors are the longitudinal degrees of freedom of optical cavities, such as Fabry-Perot cavities, recycling cavities, monolithic cavities as the output mode cleaner, etc. In the following, the lock of two cavities inherent to the work in this thesis will be described: the filter cavity in the quantum noise reduction system (see Sect. 5.3) and the new high finesse output mode cleaner cavity (see Sect. 10.3).

In order to be able to lock these cavities longitudinally, it is necessary to find a good error signal which gives us information about the length of the cavity with respect to its resonance point. A good error signal must be *bipolar*, i.e. it changes sign around the operating point; *linear*, i.e. it is proportional to the parameter we want to measure; and *cross zero* around the chosen operating point. The Pound-Drever-Hall (PDH) technique (see Sect. 2.4.2) is used to obtain an error signal that

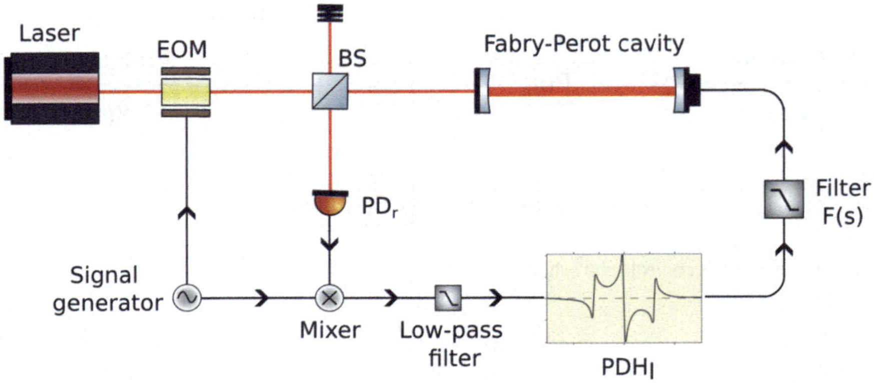

Fig. 2.14 Pound-Drever-Hall (PDH) technique scheme. The laser is modulated with an EOM, which introduces sidebands at the modulation frequency. They are reflected by the cavity since they are in anti-resonance, while the carrier beam resonates. The signal detected by the photodiode in reflection is demodulated by multiplication with the modulation frequency. A low-pass filter is applied to obtain the PDH error signal. This signal is filtered and the correction is applied to the coils of the end mirror of the cavity

meets these characteristics and it controls the length of the cavity with respect to the laser frequency [42].

The PDH scheme is shown in Fig. 2.14.

The laser, at ω-frequency, passes through an Electo-Optic Modulator (EOM) where it is frequency modulated, acquiring sidebands at a frequency $\omega \pm \Omega$, chosen to be anti-resonant with the Fabry-Perot cavity. The field exiting the EOM, E_{in} can be written as:

$$E_{in} = E_0 \cdot e^{i(\omega t + m \cdot sin(\Omega t))} \tag{2.59}$$

where $E_0 \cdot e^{i\omega t}$ is the field entering the EOM, and m is the modulation depth and Ω is the modulation frequency. Expanding the exponential in terms of Bessel functions, one obtains

$$E_{in} = E_0 \cdot \left[J_0(m) \cdot e^{i\omega t} + J_1(m) \cdot e^{i(\omega+\Omega)t} - J_1(m) \cdot e^{i(\omega-\Omega)t} \right] \tag{2.60}$$

where it has been considered that the modulation depth is small, i.e. $m < 1$, meaning that all the power is in the carrier and sidebands fields. The carrier (corresponding to the $J_0(m)$ term) must resonate in the cavity, while the sidebands (corresponding to the $J_1(m)$ terms) are reflected. The sideband shifted by $+\Omega$ with respect to the carrier is called the *upper sideband* and the sideband shifted by $-\Omega$ is called the *lower sideband*. The difference in phase between the sidebands and the carrier contains information about the length of the cavity and it is therefore the PDH error signal. The reflected field can be described as

$$E_r = E_0 e^{i\omega t} \cdot \left[r(\omega) J_0(m) + r(\omega + \Omega) J_1(m) e^{i\Omega t} - r(\omega - \Omega) J_1(m) e^{-i\Omega t} \right] \tag{2.61}$$

where $r(\omega)$, $r(\omega + \Omega)$ and $r(\omega - \Omega)$ are the reflectivities of the cavity given by Eq. 2.16 for the carrier field, upper and lower sidebands respectively. The signal that arrives on the photodiode is the reflected power, given by Black [42]

$$\begin{aligned} P_r = {} & P_c \cdot |r(\omega)|^2 + P_s \cdot \left[|r(\omega + \Omega)|^2 + |r(\omega - \Omega)|^2 \right] + \\ & + 2\sqrt{P_c P_s} \cdot \left[Re\left(r(\omega) r^*(\omega + \Omega) - r^*(\omega) r(\omega - \Omega) \right) \cos(\Omega t) + \right. \\ & \left. + Im\left(r(\omega) r^*(\omega + \Omega) - r^*(\omega) r(\omega - \Omega) \right) \sin(\Omega t) \right] + (2\Omega \quad \text{terms}). \end{aligned} \tag{2.62}$$

where $P_c = |E_0|^2 \cdot J_0^2(m)$ is the power of the carrier field and $P_s = |E_0|^2 \cdot J_1^2(m)$ is the power of the sideband fields. The first and second terms of Eq. 2.62 are the DC terms, i.e. the static power in reflection of the cavity; the third term represents the interference between the carrier and the sidebands, which is the term of interest. In order to extract the information carried by this term, it is necessary to demodulate the error signal, i.e. to multiply it with the signal used to generate the modulation frequency ($\cos(\Omega t + \phi_0)$) using a mixer, as shown in Fig. 2.14. The resulting signal contains a DC term carrying the phase information needed, and other terms in Ω and 2Ω that are filtered applying a low-pass filter. Depending on the chosen demodulation phase ϕ_0, the two quadratures of the signal can be extracted: the in-phase quadrature (I) by multiplying by $\sin(\Omega t + \phi_0)$ (PDH_I) and the quadrature by multiplying by $\cos(\Omega t + \phi_0)$ (PDH_Q). The phase ϕ_0 is adjusted so that the error signal is fully in one quadrature and zero in the other when the cavity is close to the resonance. The analytical expression of the error signal is given by Black [42]

$$\begin{aligned} PDH_I &= 4 \cdot \sqrt{P_c P_s} r_2 \cdot t_1^2 \cdot \frac{\sin(2k\delta l)}{1 + r_1^2 \cdot r_2^2 - 2 \cdot r_1 \cdot r_2 \cdot \cos(2k\delta l)} \\ PDH_Q &= 0. \end{aligned} \tag{2.63}$$

These signals are shown in Fig. 2.15. As one can see, when the cavity crosses the resonance, the error signal is linear and bipolar. The error signal PDH_I is then used in a feedback loop to control the position of the mirror of the cavity.

This technique is used to control all the Virgo cavities (FP, PRC, SRC). The modulation frequencies used for the longitudinal controls are 6, 8, and 56 MHz.

Alternatively, instead of modulating the laser frequency, the cavity length can be modulated in order to obtain the error signal. This is fully equivalent and this technique is used for the control of the OMC cavity, as explained in Chap. 10.

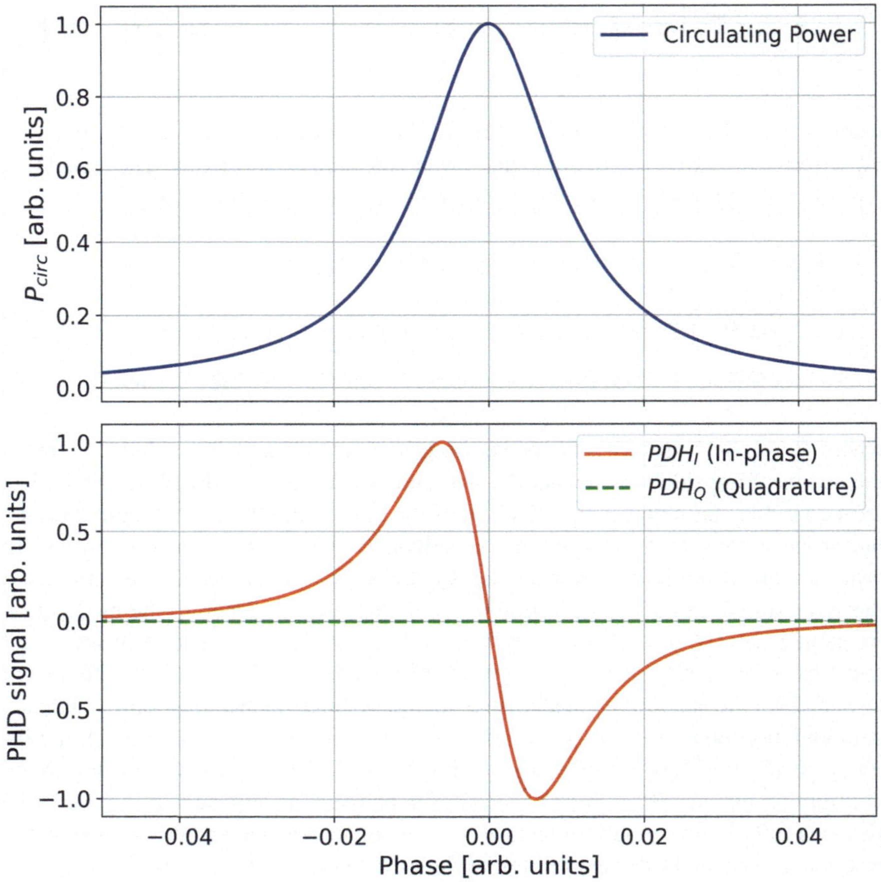

Fig. 2.15 On the top, the power circulating inside the cavity. On the bottom, the demodulated signals for the PDH in reflection: PDH_I and PDH_Q

2.5 Towards a Third-Generation of Gravitational Waves Detectors

A new era of observing the universe has begun thanks to the ground-based gravitational wave detectors LIGO and Virgo. They have so far revealed 90 gravitational signals from the most distant regions of the cosmos. But this is only the beginning of this bright astronomical era: in the coming years, thanks to their upgrades and together with two other gravitational wave detectors, KAGRA in Japan [35] and LIGO India [43], it will be possible to discover more of the mysteries hidden in our universe.

Scientists are also already looking forward to the next generation of gravitational wave detectors, the construction of which is expected to start around the 2030s. The

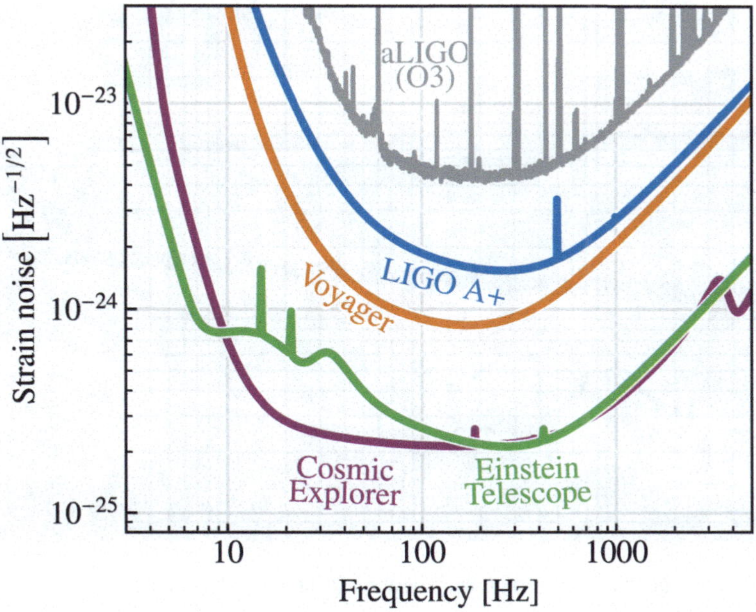

Fig. 2.16 Sensitivity of the second and third generation ground-based detectors. The different colors represent the sensitivity of the different detectors: in grey aLIGO during the run O3, in blue aLIGO+ during O4, in orange Voyager (a foreseen LIGO update), in green Einstein Telescope and in purple Cosmic Explorer [47]

space-based LISA detector [44] has been proposed to detect gravitational waves in a different frequency region than current detectors, to pick up signals between fractions of mHz and a few Hz. The third generation of ground-based detectors will consist of the Einstein Telescope detector in Europe [45] and the Cosmic Explorer in the United States [46]. The sensitivity of the different detectors is shown in Fig. 2.16. The expected number of coalescence events that will be detected is huge and for ET the detection rate for coalescence events [48] is

- $10^5 - 10^6$ BBH per year;
- $7 \cdot 10^4$ BNS per year;
- $\sim 100 - 1000$ BNS with electro-magnetic counterpart per year.

These new detectors could provide answers about the role of dark matter and dark energy in the universe. Thanks to the huge number of coalescences of black holes, it will be possible to study the physics of black holes in greater detail and to enable new theories of modified gravity to be tested. In addition, it will be possible to detect a large number of coalescence events of neutron stars, allowing to understand the behavior of matter under conditions of extreme density and pressure (which cannot be observed in th laboratory) and broaden the horizons of multi-messenger astronomy.

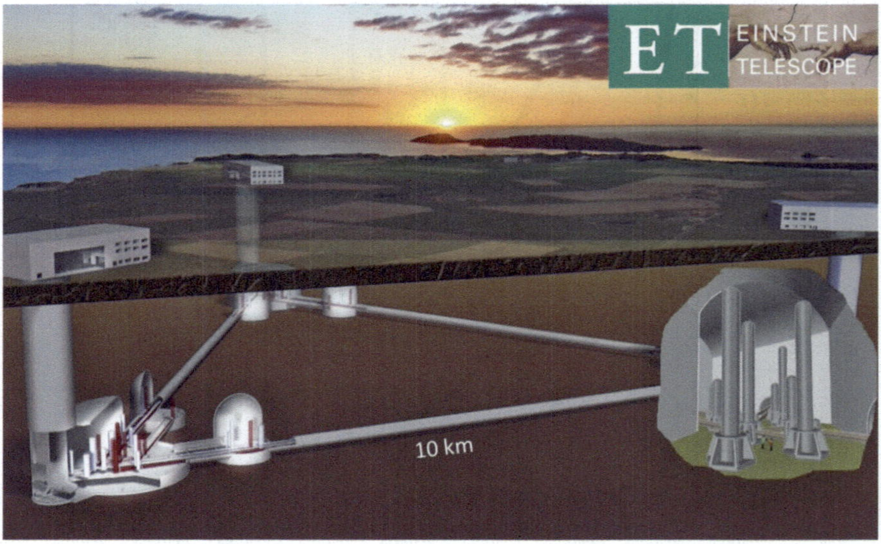

Fig. 2.17 Einstein Telescope infrastructure design. Credit: ET collaboration

2.5.1 Einstein Telescope

The Einstein Telescope (ET) detector will be built underground to attenuate seismic noise, at a depth of 200–300 m, like KAGRA. It will consist of three nested detectors, each consisting of two interferometers, each with a dual-recycled Michelson layout. Its geometry will be triangular, as shown in Fig. 2.17 with cavities along the arms 10 km long. Three of these detectors will be used for low frequencies and the others for high frequencies. The mirrors of the interferometers operating at low frequencies will be at cryogenics temperature in order to decrease the thermal noise. Frequency dependent squeezing will be used for quantum noise reduction.

2.6 Conclusion

In this chapter the principle used to detect gravitational wave signals, namely laser interferometry, has been described. The different sources of noises that limit the sensitivity of the gravitational waves detectors have been introduced. Among them is the quantum noise, given by the shot noise which deteriorates the sensitivity at high frequency and by the radiation pressure noise which worsen the low frequency. A technique, known as squeezing, has been implemented to reduce the shot noise during the last observing run (O3), increasing the radiation pressure noise. In the next chapter we will describe how it is possible to reduce both contributions at the same time.

Part II
Broadband Quantum Noise Reduction in Advanced Virgo Plus

Chapter 3
Theory of Quantum Light in Gravitational Waves Detectors

Quantum noise limits the sensitivity of both high-frequency and low-frequency gravitational wave detectors. This results from the quantum nature of the light arriving at the detector and mirrors. In this Chapter, the theoretical treatment of the quantization of light is described as well as its different states: Fock states, coherent states and squeezed states. In addition, the quantum theoretical description of an interferometer and of quantum noise is given, concluding explaining how squeezed states can be used to reduce the quantum noise.

3.1 Quantization of the Radiation Field

In vacuum electromagnetic radiation is described from Maxwell's equations by the vector potential $\mathbf{A}$, having chosen the gauge $\nabla \cdot \mathbf{A} = 0$, as

$$-\nabla^2 \mathbf{A} + \frac{1}{c^2}\frac{\partial^2 \mathbf{A}}{\partial t^2} = 0. \tag{3.1}$$

The quantization of the electromagnetic field proceeds by the replacement of the classical vector potential $\mathbf{A}$ by a quantum-mechanical operator $\hat{A}$. The classical theory can be expressed in a form from which the transition to quantum mechanics is straightforward. The electric field of an electromagnetic wave in free space in $r = (x, y, z)$ at time t is

$$\hat{E}(r,t) = i \sum_k \sqrt{\frac{\hbar\omega_k}{2\epsilon_0}} \left[\hat{a}_k u_k(r) e^{-i\omega_k t} + \hat{a}_k^\dagger u_k^*(r) e^{i\omega_k t}\right] \tag{3.2}$$

where ϵ_0 is the vacuum permittivity and the sum over k runs over different modes of the electric field with angular frequencies ω_k and spatial distribution functions

E. Polini, *Broadband Quantum Noise Reduction in Advanced Virgo Plus*,
Springer Theses, https://doi.org/10.1007/978-3-031-95143-5_3

$u_k(r)$. These functions are an orthonormal set which describe spatial modes properties such as propagation direction, mode shape, and polarization. In a classical description, $\hat{a}_k$ and $\hat{a}^\dagger$ represent complex numbers describing dimensionless field amplitude, while in a quantum-mechanical approach they are replaced by operators. These operators follow the commutation rules

$$\left[\hat{a}_k, \hat{a}_{k'}\right] = \left[\hat{a}_k^\dagger, \hat{a}_{k'}^\dagger\right] = 0 \qquad \left[\hat{a}_k, \hat{a}_{k'}^\dagger\right] = \delta_{kk'}. \tag{3.3}$$

The Hamiltonian for the electromagnetic field, treated as an harmonic oscillator, can be written as

$$\hat{H} = \sum_k \hbar\omega_k \left(\hat{a}_k^\dagger \hat{a}_k + \frac{1}{2}\right) \tag{3.4}$$

where $\hat{a}_k$ and $\hat{a}^\dagger$ that are ladder operators for quantum harmonic oscillator, represent here the creation and annihilation operator of a photon with wave vector k. It is clear that these operators are not Hermitian and that means they cannot represent observable quantities. Their properties of destroying or creating a quantum of energy is, however, easily appreciated in a physical sense. It is convenient to introduce the *quadrature operators*

$$\hat{X}_1 = \frac{1}{2}\left(\hat{a}^\dagger + \hat{a}\right) \qquad \hat{X}_2 = \frac{1}{2}i\left(\hat{a}^\dagger - \hat{a}\right) \tag{3.5}$$

with reverse relations

$$\hat{a} = \hat{X}_1 + i\hat{X}_2 \qquad \hat{a}^\dagger = \hat{X}_1 - i\hat{X}_2 \tag{3.6}$$

in order to obtain an Hermitian operator condition between them. It is easy to show that the Hamiltonian of Eq. 3.4 can be expressed as (fixing k)

$$\hat{H} = \hbar\omega\left(\hat{X}_1^2 + \hat{X}_2^2\right) \tag{3.7}$$

and that the quadrature-operator commutation relation is

$$\left[\hat{X}_1, \hat{X}_2\right] = \frac{i}{2}. \tag{3.8}$$

The Heisenberg position-momentum uncertainty relation can be expressed as

$$\left(\Delta\hat{X}_1\right)^2 \left(\Delta\hat{X}_2\right)^2 \geq \frac{1}{16}. \tag{3.9}$$

Considering only the time propagation of the scalar electromagnetic field, the field can be re-written as

$$\hat{E}(t) = E_0 \left(\hat{a} e^{-i\omega t} + \hat{a}^{\dagger} e^{i\omega t} \right) \tag{3.10}$$

from Eq. 3.2. Using quadrature operators, it can be expressed as

$$\hat{E}(t) = 2E_0 \left[\left(\hat{X}_1 + \Delta \hat{X}_1 \right) \cos(\omega t) + \left(\hat{X}_2 + \Delta \hat{X}_2 \right) \sin(\omega t) \right]. \tag{3.11}$$

where $\hat{X}_1$ and $\hat{X}_2$ correspond to field amplitudes of oscillations with a phase difference of 90° and $\Delta \hat{X}_1$ and $\Delta \hat{X}_2$ are the field fluctuations in the quadrature space.

3.1.1 Fock States

The *photon number operator* is defined as

$$\hat{n}_k := \hat{a}_k^{\dagger} \hat{a}_k \tag{3.12}$$

and its eigenstates, written as $|n_k\rangle$, are called the *number states* or *Fock states*. They satisfy the following relation

$$\hat{n}_k |n_k\rangle = n_k |n_k\rangle \tag{3.13}$$

and form a complete set of basis vectors for a Hilbert space. The vacuum state, which is the ground state of the harmonic oscillator, is defined by

$$\hat{n}_k |0\rangle = 0. \tag{3.14}$$

The action of the operators $\hat{a}_k$ and $\hat{a}^{\dagger}$ on a Fock state is given by

$$\hat{a}_k |n_k\rangle = \sqrt{n_k} |n_k - 1\rangle \tag{3.15}$$

$$\hat{a}_k^{\dagger} |n_k\rangle = \sqrt{n_k + 1} |n_k + 1\rangle \tag{3.16}$$

where one photon with wave vector k is annihilated and created respectively in Eqs. 3.15 and 3.16. Photon number states are a useful representation of optical fields with very low photon numbers. In the following the index k will be omitted, since only a single optical mode is considered.

3.1.2 Coherent States

The continuous-wave laser beams are well described by *coherent states* $|\alpha\rangle$, where α is a complex number. In contrast to Fock states, where the photon number is know accurately but the phase remains unknown, knowledge about both the amplitude and phase of a coherent state can be obtained. However, the Heisenberg uncertainty principle puts an upper bound on the accuracy of this knowledge. The generation of a coherent state is described by the *displacement operator* $\hat{D}(\alpha)$ acting on the vacuum state

$$|\alpha\rangle := \hat{D}(\alpha)|0\rangle \tag{3.17}$$

with

$$\hat{D}(\alpha) := exp(\alpha\hat{a} - \alpha^*\hat{a}^\dagger), \qquad \alpha \in \mathbb{C} \tag{3.18}$$

where $\alpha = |\alpha|e^{i\theta}$. Coherent states are eigenstates of the annihilation and creation operators $\hat{a}$ and $\hat{a}^\dagger$

$$\hat{a}|\alpha\rangle = \alpha|\alpha\rangle \tag{3.19}$$

$$\langle\alpha|\,\hat{a}^\dagger = \langle\alpha|\,\alpha^*. \tag{3.20}$$

The annihilation and creation operators are non-Hermitian operators.[1] It is useful to express coherent states as a function of number states as

$$|\alpha\rangle = e^{-|\alpha|^2/2}\sum_{n=0}^{\infty}\frac{\alpha^n}{\sqrt{n!}}|n\rangle. \tag{3.21}$$

The projection of a coherent state in an eigenstate of the number operator $|n\rangle$ is

$$\langle n|\alpha\rangle = \frac{\alpha^n}{\sqrt{n!}}e^{-|\alpha|^2/2} \tag{3.22}$$

and the mean value of the photon number $\overline{n}$ of the coherent state $|\alpha\rangle$ is

$$\overline{n} = \langle n\rangle = \langle\alpha|n|\alpha\rangle = \langle\alpha|\hat{a}^\dagger\hat{a}|\alpha\rangle = \langle\alpha|\alpha^*\alpha|\alpha\rangle = |\alpha|^2\,\langle\alpha|\alpha\rangle = |\alpha|^2. \tag{3.23}$$

[1] An operator $\hat{O}$ is Hermitian if $\hat{O} = \hat{O}^\dagger$, so if its eigenvalues are real, and our is not the case since α is a complex number.

The probability distribution function of the photons in coherent state is Poissonian and given by

$$P(n) = |\langle n|\alpha\rangle|^2 = e^{-|\alpha|^2}\frac{|\alpha|^{2n}}{n!}. \tag{3.24}$$

For a coherent state $|\alpha\rangle$ the expectation values of the two quadrature operators are the real and imaginary part of the complex field amplitude α

$$\langle\alpha|\hat{X}_1|\alpha\rangle = Re(\alpha), \qquad \langle\alpha|\hat{X}_2|\alpha\rangle = Im(\alpha). \tag{3.25}$$

The $\hat{X}_1$ quadrature is conventionally called the amplitude quadrature and $\hat{X}_2$ the phase quadrature. The quadrature operator along an arbitrary angle ϕ are defined as

$$\hat{X}_\phi := \hat{X}_1 \cos\phi + \hat{X}_2 \sin\phi = \frac{1}{2}\left(\hat{a}^\dagger e^{i\phi} + \hat{a}e^{-i\phi}\right). \tag{3.26}$$

From Heisenberg uncertainty principle and the commutation rules, it is obtained that

$$\left(\Delta\hat{X}_1\right)^2\left(\Delta\hat{X}_2\right)^2 = \frac{1}{16} \tag{3.27}$$

and the variance for the two different quadratures are

$$\Delta\hat{X}_1 = \Delta\hat{X}_2 = \frac{1}{4}. \tag{3.28}$$

From Eq. 3.27, coherent states belong to the class of minimum-uncertainty states. The laser light can be described by quantum states of coherent light with their respective fluctuations, according to Eq. 3.11 and it can be represented as in Fig. 3.1.

Furthermore, when the field amplitude is zero, the fluctuations remain. Actually, the quantum coherent vacuum can be represented as in Fig. 3.2.

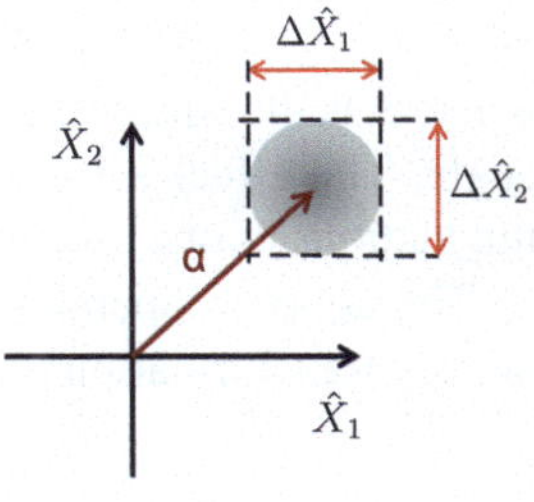

Fig. 3.1 Bright coherent state of light: $|\alpha\rangle$

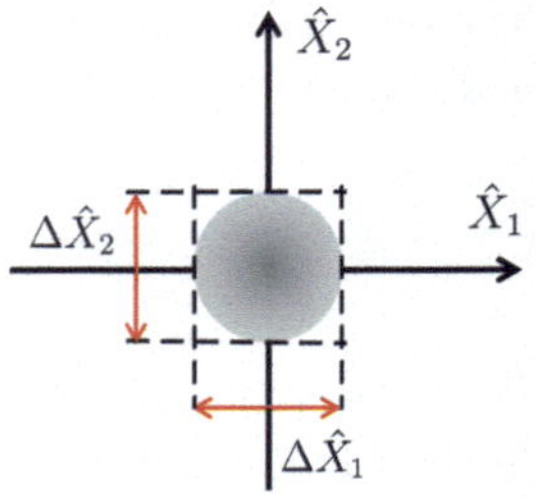

Fig. 3.2 Coherent vacuum state: $|0\rangle$

3.1.3 Squeezed States

The *squeezed states* are a special class of minimum uncertainty states. The Heisenberg uncertainty principle presented in the Eq. 3.27 places a relation on the product of the variances of the two quadratures but not on the individual variances. The fluctuations of a quadrature can therefore be lower than those of a coherent state, *squeezed quadrature*,

$$\Delta\hat{X}_1 < \frac{1}{4} \tag{3.29}$$

while those in orthogonal quadrature are greater, *anti-squeezed quadrature*,

$$\Delta\hat{X}_2 > \frac{1}{4}. \tag{3.30}$$

The generation of a *squeezed vacuum state* is described by applying the *squeeze operator* $\hat{S}(\xi)$ to the vacuum state

$$|\xi\rangle := \hat{S}(\xi)|0\rangle \tag{3.31}$$

with

$$\hat{S}(\xi) := exp\left(\frac{1}{2}\xi^*\hat{a}^2 - \frac{1}{2}\xi\hat{a}^{\dagger 2}\right), \qquad \xi = re^{2i\theta} \tag{3.32}$$

where ξ is the squeezing parameter, r is the squeezing factor and θ is the squeezing angle. The effect achieved on the vacuum state $|0\rangle$ is to reduce, i.e. squeeze, the fluctuations in one quadrature direction, as shown on the left side of Fig. 3.3.

The squeeze operator is unitary $\hat{S}^\dagger(\xi) = \hat{S}^{-1}(\xi) = \hat{S}(-\xi)$. The most useful unitary transformation properties of this operator are

$$\hat{S}^\dagger(\xi)\hat{a}\hat{S}(\xi) = \hat{a}\cosh r - \hat{a}^\dagger e^{i2\theta}\sinh r \tag{3.33}$$

$$\hat{S}^\dagger(\xi)\hat{a}^\dagger\hat{S}^(\xi) = \hat{a}^\dagger\cosh r - \hat{a}e^{-i2\theta}\sinh r. \tag{3.34}$$

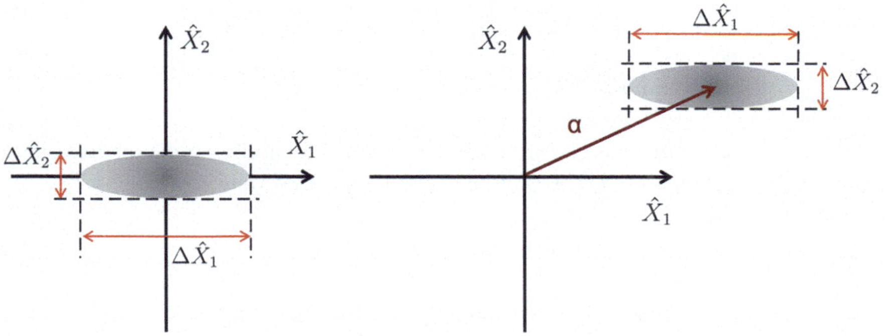

Fig. 3.3 On the left, a *squeezed vacuum state* with $\theta = 0$: $|\xi\rangle$. On the right, a *squeezed coherent state* with $\theta = 0$: $|\alpha, \xi\rangle$. These states are squeezed along the $\hat{X}_2$ direction

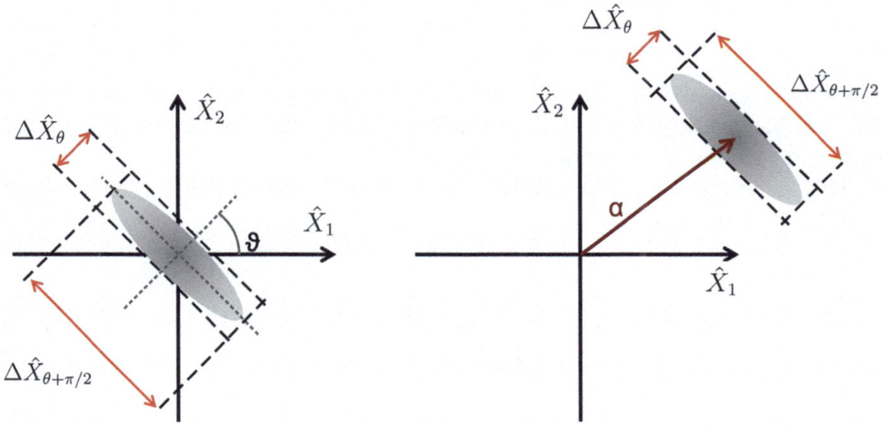

Fig. 3.4 On the left, a squeezed vacuum state with $\theta \neq 0$: $|\xi\rangle$. On the right, a squeezed coherent state with $\theta \neq 0$: $|\alpha, \xi\rangle$. These states are squeezed along the $\hat{X}_\theta$ direction

The squeezed vacuum states can be rotated in the space of quadratures by an angle θ, as shown on the left side of Fig. 3.4.

The variances for these states are obtained from Eqs. 3.33 and 3.34

$$\left(\Delta\hat{X}_\theta\right)^2 = \frac{1}{4}e^{-2r}, \qquad \left(\Delta\hat{X}_{\theta+\pi/2}\right)^2 = \frac{1}{4}e^{2r} \tag{3.35}$$

where θ corresponds to the squeezing quadrature and $\theta + \pi/2$ to the anti-squeezing one. The variances express the compression factor of these states, which corresponds to how much the fluctuations are reduced in one direction and increased in the perpendicular direction. For a general quadrature angle ϕ that is not aligned with the squeezing angle θ the variance becomes

$$\left(\Delta\hat{X}_\phi\right)^2 = \frac{1}{2}\left(\cos^2(\phi-\theta)e^{-2r} + \sin^2(\phi-\theta)e^{2r}\right) = \\ = \cosh(2r) - \sinh(2r)\cos[2(\theta-\phi)]. \tag{3.36}$$

Moreover, a *bright squeezed state* can be obtained by applying the displacement operator $\hat{D}(\alpha)$, defined in Eq. 3.18, to a vacuum squeezed state

$$|\alpha, \xi\rangle := \hat{D}(\alpha)|\xi\rangle = \hat{D}(\alpha)\hat{S}(\xi)|0\rangle. \tag{3.37}$$

The effect on the vacuum squeezed state $|\xi\rangle$ is to displace it by $|\alpha$ in the quadrature space, as shown on the right side of Fig. 3.3. This means that a squeezed state is a bright state, so it contains photons. The expected photon number of a bright squeezed state is

$$\overline{n} = \langle n\rangle = \langle\alpha, \xi|\hat{n}|\alpha, \xi\rangle = |\alpha|^2 + \sinh^2 r. \tag{3.38}$$

These states can also be rotated by an angle θ in the quadrature space as shown in the right-hand side of Fig. 3.4. The variances of these states are the same as for the squeezed vacuum, as shown in Eqs. 3.35 and 3.36.

The squeezing level is generally given in decibel (dB) which can be obtained from Eq. 3.35 by the relationship

$$V[dB] = 10\log\left(\frac{\left(\Delta\hat{X}_\theta\right)^2}{\left(\Delta\hat{X}_{\theta+\pi/2}\right)^2}\right) = 20\log\left(e^{-2r}\right). \tag{3.39}$$

3.2 Generation of Squeezed States

Vacuum squeezed states and bright squeezed states can be generated by a non linear process in dielectric medium, the *optical parametric amplification*. A medium is said to be nonlinear when the polarization density $P(t)$ responds nonlinearly to an applied electric field $E(t)$ [49] as

$$P(t) = \epsilon_0\left[\chi^{(1)}E(t) + \chi^{(2)}E(t)^2 + \chi^{(3)}E(t)^3 + \ldots\right] \tag{3.40}$$

where ϵ_0 is the permittivity of the vacuum and $\chi^{(n)}$ is the n-th order of the optical susceptibility, with typical values of $\chi^{(1)} \sim 1$, $\chi^{(2)} \sim 10^{-12}\frac{m}{V}$ and $\chi^{(2)} \sim 10^{-12}\frac{m^2}{V^2}$ for the usual nonlinear solid state optical material. For this optical process, susceptibilities of order greater than two, can be neglected. The focus of this section concerns the generation of vacuum squeezed states, since they will be largely used in the following sections. The non-linear crystal is pumped with

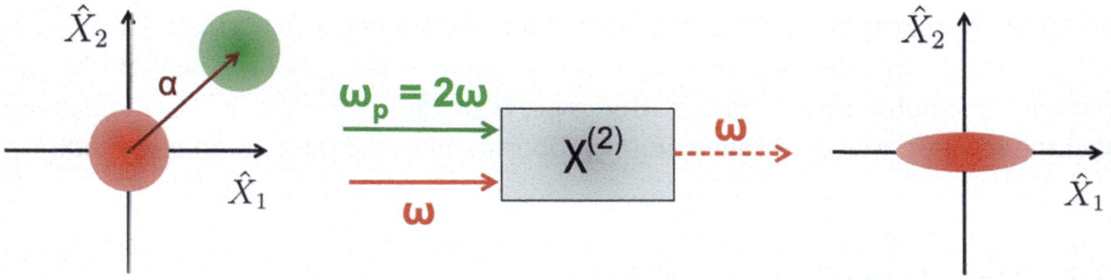

Fig. 3.5 Production of squeezed vacuum states with a non-linear degenerate medium. The crystal is pumped with a green beam at ω_p interacting with the coherent vacuum at ω. The produced state is a squeezed vacuum state at the same frequency ω

an electromagnetic field of pulsation ω_p which introduces a separation of charges and thus a nonlinear dielectric polarization. Vacuum fluctuations, shown in Fig. 3.2, enter the crystal and the ones with angular frequency $\omega = \omega_p/2$ are transformed into squeezed fluctuations. The optical parametric amplification is shown in Fig. 3.5.

The interaction between the vacuum fluctuation field and the pump field inside the non linear crystal, results in a phase-dependent amplification and de-amplification of the quantum uncertainty at the frequency ω. The stronger the pump field, the stronger the parametric amplification will be. The object in which this process take place is called *Optical Parametric Oscillator (OPO)*. More details on the graphical description of the optical parametric generation of squeezed states can be found in [50].

This phenomenon can be treated with a quantum formalism by considering the Hamiltonian of interaction between the vacuum field and the pump field inside the OPO [51], defined as

$$\hat{\mathcal{H}} = \hbar\omega\hat{a}^\dagger\hat{a} + \hbar\omega_p\hat{b}^\dagger\hat{b} + i\hbar\chi^{(2)}\left(\hat{a}^2\hat{b}^\dagger - \hat{a}^{\dagger 2}\hat{b}\right) \tag{3.41}$$

where $\hat{a}$ and $\hat{b}$ are the annihilation operators of the vacuum field (ω) and the pump field (ω_p) respectively. The operator $\hat{b}$ can be expressed as $\hat{b} = \beta e^{-i\omega_p t}$ and the interaction term, the third term of Eq. 3.41, becomes

$$\hat{\mathcal{H}}_I(t) = i\hbar\chi^{(2)}\left[\beta^*\hat{a}^2 e^{i(\omega_p - 2\omega)t} - \beta\hat{a}^{\dagger 2} e^{-i(\omega_p - 2\omega)t}\right] \tag{3.42}$$

that generally is time dependent. Considering the degenerate process where $\omega_p = 2\omega$, Eq. 3.42 becomes

$$\hat{\mathcal{H}}_I = i\hbar\chi^{(2)}\left(\beta^*\hat{a}^2 - \beta\hat{a}^{\dagger 2}\right) \tag{3.43}$$

that is time-independent. The evolution operator associated to the interaction Hamiltonian is

$$\hat{U}_I(t, 0) = e^{-i\frac{\hat{\mathcal{H}}_I t}{\hbar}} = e^{\chi^{(2)}(\beta^*\hat{a}^2 t - \beta\hat{a}^{\dagger 2} t)} \tag{3.44}$$

and it has the form of the squeeze operator introduced in Eq. 3.32, $\hat{U}_I(t, 0) = \hat{S}(\xi)$, for $\xi = 2\chi^{(2)}\beta t$, showing that squeezed vacuum states can be generated by the parametric amplification process. Moreover, being $\xi = 2\chi^{(2)}\beta t$, it is demonstrated that the level of produced squeezing is proportional to the pump field amplitude.

3.3 Detection of Squeezed States

To be able to characterize squeezed states, it is necessary to measure fluctuations in both quadratures. For this purpose, an homodyne detector is used, as illustrated in [52]. Its scheme is shown in Fig. 3.6. The beam to be measured recombines with a reference beam, called *local oscillator (LO)*, which has the same frequency and tunable phase shift ϕ, with a 50/50 beam splitter (balanced detection). The two fields entering the beam splitter can be expressed, using a linear approximation, as

$$\hat{a} = \alpha + \delta\hat{a} \tag{3.45}$$

for the squeezed field ($\alpha = 0$ for a vacuum state), and

$$\hat{b} = (\beta + \delta\hat{b})e^{i\phi} \tag{3.46}$$

for the LO field, where $\delta\hat{a}$ and $\delta\hat{b}$ are the continuum of fluctuating modes surrounding the carriers α and β respectively.

The two output fields can be described using the so-called *beam splitter model*

$$\begin{pmatrix} \hat{c} \\ \hat{d} \end{pmatrix} = \begin{pmatrix} r & t \\ t & -r \end{pmatrix} \begin{pmatrix} \hat{a} \\ \hat{b} \end{pmatrix} \tag{3.47}$$

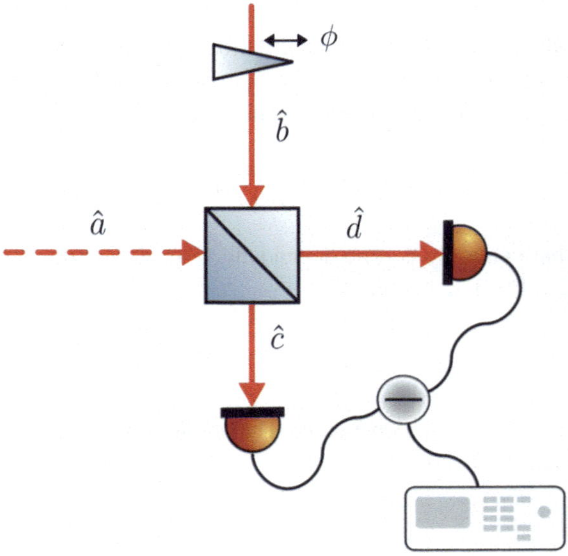

Fig. 3.6 Balanced homodyne detector scheme. The local oscillator beam passes through a phase modulator acquiring a modulation phase ϕ. The squeezed beam $\hat{a}$ recombines with the local oscillator beam $\hat{b}(\phi)$ on a 50/50 beam splitter. The output fields $\hat{c}$ and $\hat{d}$ arrive on two photodetectors. The beat-note between the power of the output fields is then read by a spectrum analyzer

being r the reflectivity and t the transmissivity of the beam splitter. Considering a 50/50 beam splitter, meaning $r = t = \frac{1}{\sqrt{2}}$, one obtains

$$\hat{c} = \frac{1}{\sqrt{2}}(\hat{a} + \hat{b}) \tag{3.48}$$

and

$$\hat{d} = \frac{1}{\sqrt{2}}(\hat{a} - \hat{b}). \tag{3.49}$$

and the associated photo-currents detected by two different photodiodes are $i_c \propto \hat{c}\hat{c}^\dagger$ and $i_d \propto \hat{d}\hat{d}^\dagger$.

Let's define the beam quadratures as

$$\delta\hat{X}_1^a = \delta\hat{a}^\dagger + \delta\hat{a} \qquad \delta\hat{X}_2^a = i(\delta\hat{a}^\dagger - \delta\hat{a}) \tag{3.50}$$

and

$$\delta\hat{X}_1^b = \delta\hat{b}^\dagger + \delta\hat{b} \qquad \delta\hat{X}_2^b = i(\delta\hat{b}^\dagger - \delta\hat{b}) \tag{3.51}$$

neglecting $O(\delta^2)$ terms. Considering that the local oscillator is much brighter than the squeezed beam ($|\alpha|^2 << |\beta|^2$), the differential output current is

$$i_- \propto i_c - i_d \propto \beta(\delta\hat{X}_1^a \cos\phi + \delta\hat{X}_2^a \sin\phi) = \beta\delta\hat{X}_\phi^a. \tag{3.52}$$

The squeezed field fluctuation in the quadrature $\hat{X}_\phi^a$ is amplified proportionally to the LO amplitude and the LO noise is completely suppressed. By means of the phase-shifter device along the LO path, it is therefore possible to detect the uncertainty of an arbitrary quadrature $\delta\hat{X}_\phi^a$.

3.4 Squeezing Degradation Mechanisms

Squeezed states are subject to several degradation mechanisms that limit the experimentally achievable quantum noise reduction [53]. These can be divided into two main categories: *optical losses* and *phase noise*. Optical losses reduce the squeezing of the squeezing quadrature and phase noise, on the other hand, introduces jitter on the squeezing ellipse which couples the two quadratures. Other sources of degradation, less limiting than those just mentioned, are also present. These include classical noise related to the electronics used to detect the signal, known as *dark noise* [54]. In addition, squeezed states are very sensitive to scattered light. The recombination of the scattered light produced within the optical system with the squeezed field, degrades these states [55]. More details will be given in Part III.

3.4.1 Optical Losses

Optical losses in the propagation of a quantum state can be described with the beam splitter model introduced in Eq. 3.47. The input fields are a coherent vacuum field $|0\rangle$ and the squeezed vacuum field $|\xi\rangle$ and they recouple inside a beam splitter of reflectivity $\sqrt{L}$ and transmissivity $\sqrt{1-L}$. The resulting output fields squeezed states are therefore less squeezed than the input field, as shown in Fig. 3.7.

The output field $\hat{d}$ can be written as

$$\hat{d} = \sqrt{1-L}\cdot\hat{a} - \sqrt{L}\cdot\hat{b} \tag{3.53}$$

where $\hat{a}$ is the input vacuum squeezed field and $\hat{b}$ is the vacuum state. The quadrature variance of this output field is

$$\left(\Delta\hat{X}_d^L\right)^2 = (1-L)\left(\Delta\hat{X}_a\right)^2 + L\left(\Delta\hat{X}_b\right)^2 \tag{3.54}$$

considering that the fluctuations of $\hat{b}$ are uncorrelated to those of $\hat{a}$. The squeezed state variance $\left(\Delta\hat{X}_a\right)^2$ can be expressed as V_- and V_+ for the squeezing and anti-squeezing quadratures respectively, while a vacuum state has variance V_{vac}.

Equation 3.54 can be therefore re-written as

$$V_\pm^L = (1-L)V_\pm + LV_{vac} \tag{3.55}$$

or also using the squeezing ratios

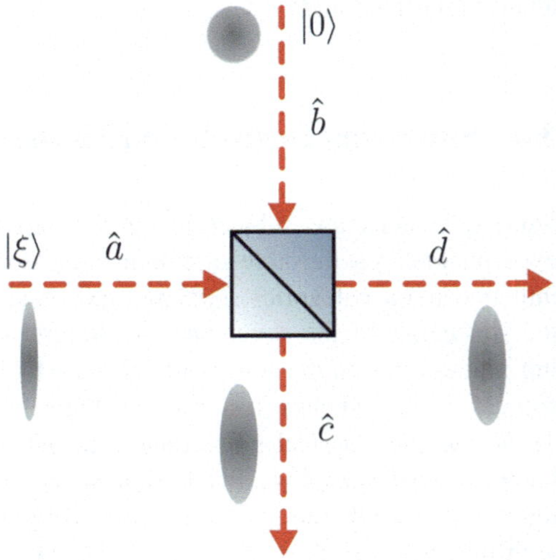

Fig. 3.7 Optical losses experienced by a vacuum squeezed state using the beam splitter model, with $r = \sqrt{L}$

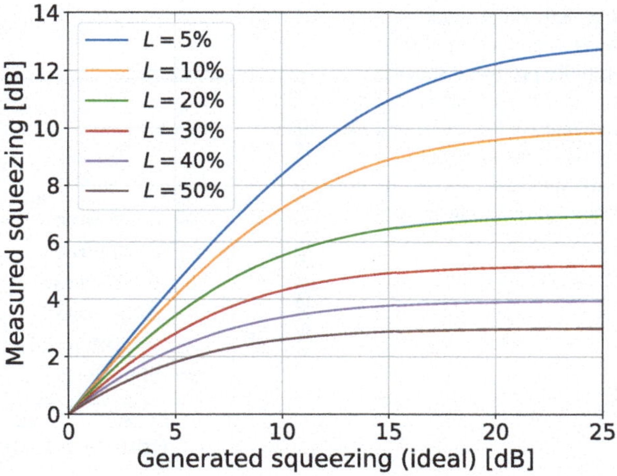

Fig. 3.8 Absolute value of squeezed level after the propagation and measurement for different values of optical losses, for an initially produced squeezing level up to 40 dB

$$R^L_{\pm} = \frac{V^L_{\pm}}{V_{vac}} = (1 - L)R_{\pm} + L. \tag{3.56}$$

The reduction of the squeezing level due to losses is illustrated in Fig. 3.8 where the squeezing level is defined using Eq. 3.56 and is expressed in dB using Eq. 3.39.

3.4.2 *Phase Noise*

Homodyne detection is used to detect squeezed states, as seen in Sect. 3.3, and the quadrature that is detected is fixed by the relative phase between the local oscillator and the squeezed field. The local oscillator is represented by the ITF beam when the squeezing is injected in Virgo. The relative phase is controlled by a feedback loop, but unfortunately there are fluctuations in this phase that mix the measured quadrature with its orthogonal quadrature. These fluctuations can be modeled as normally distributed around $\overline{\theta} = 0$, with a root-mean-square (RMS) deviation $\tilde{\theta}$ and the variance of the resulting measurement can be written as given in [56]

$$V^{\tilde{\theta}}_{\pm} \sim V_{\pm} \cos^2 \tilde{\theta} + V_{\mp} \sin^2 \tilde{\theta}. \tag{3.57}$$

The different phase noise sources are described in [57]. The degradation of the squeezing level due to phase noise is illustrated in Fig. 3.9 where the squeezing level is defined using Eq. 3.56 and is expressed in dB using Eq. 3.39. Squeezing is indicated with a negative scale, i.e. below shot noise and anti-squeezing above. Squeezed states in the presence of much phase noise are converted to anti-squeezing.

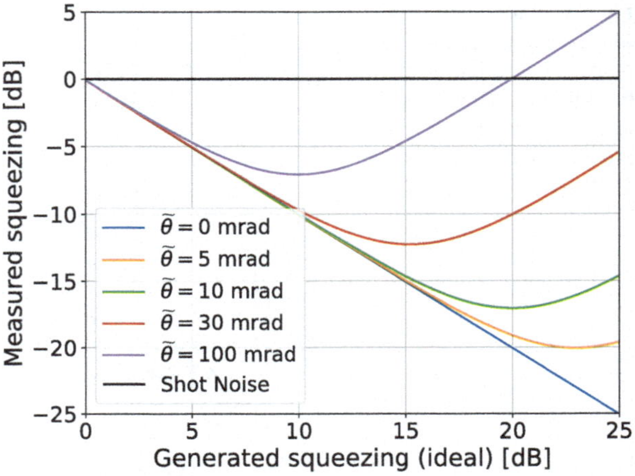

Fig. 3.9 Final squeezing level as a function of the initially produced squeezing (up to 40 dB). The squeezing, indicated below the shot noise level, is converted into anti-squeezing, above the shot noise, for high phase noise

3.4.3 *Total Degradation*

The combined effect of optical losses and phase noise on a squeezed state, can be written, from Eqs. 3.56 and 3.57, [58]

$$R_{\pm}^{L,\tilde{\theta}} = (1 - L)\left(R_{\pm}\cos^2\tilde{\theta} + R_{\mp}\sin^2\tilde{\theta}\right) + L. \tag{3.58}$$

The effect of the combination of optical losses and phase noise, as described in Eq. 3.58, is shown in Fig. 3.10.

3.5 Two-Photon Formalism

The squeezed states are generated by a particular device, the parametric amplifier as seen in Sect. 3.2, in which photons in the output modes are created or destroyed two at a time. It is therefore needed a new formalism to describe them, know as the *two-photon formalism*, introduced by Caves and Schumaker in 1985 [59, 60].

It is useful to start with the simple case of the classical field interpreted as a continuous-mode field. A laser beam, as seen in the previous section, can be described by a classical carrier at frequency ω_0 surrounded by quantum fluctuations. Vacuum states, on the other hand, have classical carriers equal to zero and have only the fluctuations. Following [58], the continuous-mode electromagnetic field can be expressed (from Eq. 3.2 and passing from discrete to continuous variables) as

$$\hat{E}(r, t) = u(r)\left[\hat{E}^{+}(t) + \hat{E}^{-}(t)\right] \tag{3.59}$$

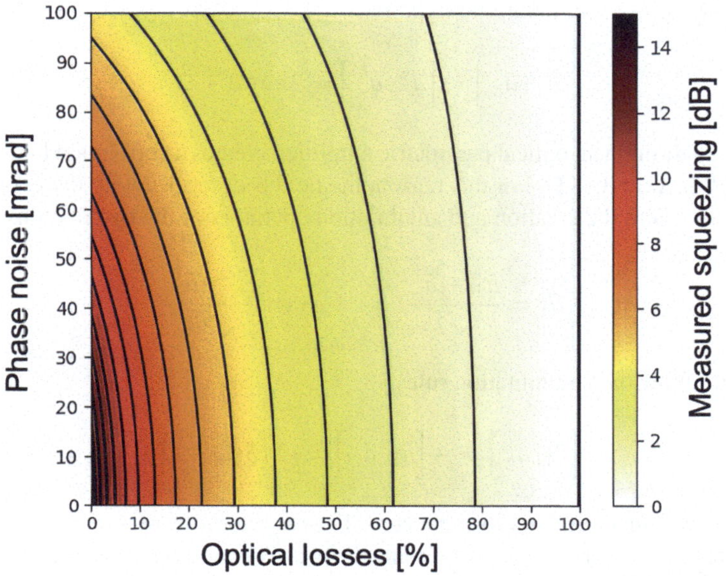

Fig. 3.10 Squeezing level in dB measured as a function of optical losses (horizontal axis) and phase noise (vertical axis), for an initially produced squeezing level of 14 dB

with

$$\hat{E}^{+}(t) = \sqrt{\frac{2\pi\hbar\omega_0}{Ac}} \int_0^{\infty} \left(\hat{a}_{\omega} e^{-i\omega t}\right) \frac{d\omega}{2\pi}, \qquad \hat{E}^{-}(t) = \left[\hat{E}^{+}(t)\right]^{\dagger} \tag{3.60}$$

where A is the effective area of the beam and $\hat{a}_{\omega}$ is the annihilation operator of each mode, which verifies the following commutation relations

$$\left[\hat{a}_{\omega}, \hat{a}_{\omega'}\right] = \left[\hat{a}^{\dagger}_{\omega}, \hat{a}^{\dagger}_{\omega'}\right] = 0 \qquad \left[\hat{a}_{\omega}, \hat{a}^{\dagger}_{\omega'}\right] = 2\pi\delta(\omega - \omega'). \tag{3.61}$$

Since the amplitude and phase fluctuations of the electromagnetic field can be considered as a modulation of the amplitude and phase respectively, it is useful to express the field as a sum of sidebands at distance Ω from the carrier (with $\Omega << \omega_0$)

$$\hat{E}(t) = \sqrt{\frac{2\pi\hbar\omega_0}{Ac}} e^{-i\omega_0 t} \int_0^{\infty} \left(\hat{a}_{+} e^{-i\Omega t} + \hat{a}_{-} e^{i\Omega t}\right) \frac{d\Omega}{2\pi} + c.c. \tag{3.62}$$

where $\hat{a}_{-}$ and $\hat{a}_{-}$ are the sideband operators defined as

$$\hat{a}_{+} = \hat{a}_{\omega_0 + \Omega} \qquad \hat{a}_{-} = \hat{a}_{\omega_0 - \Omega} \tag{3.63}$$

and following the commutation relations

$$\left[\hat{a}_+, \hat{a}_+^{\dagger'}\right] = \left[\hat{a}_-, \hat{a}_-^{\dagger'}\right] = 2\pi\delta(\Omega - \Omega'). \tag{3.64}$$

Passing through an optical parametric amplifier, creates a correlation between the sidebands at Ω and $-\Omega$. For this reason, the field is described with the two-photon formalism where the creation and annihilation operators of the two photons are

$$\hat{a}_1 = \frac{\hat{a}_+ + \hat{a}_-^\dagger}{\sqrt{2}} \qquad \hat{a}_2 = \frac{\hat{a}_+ - \hat{a}_-^\dagger}{\sqrt{2}i} \tag{3.65}$$

and they follow the commutation rules

$$\left[\hat{a}_1, \hat{a}_{2'}^\dagger\right] = -\left[\hat{a}_2, \hat{a}_{1'}^\dagger\right] = i2\pi\delta(\Omega - \Omega') \tag{3.66}$$

where all the other combinations are zero. The operator $\hat{a}_1$ represents the amplitude quadrature and the operator $\hat{a}_2$ represents the phase quadrature. The electromagnetic field can be re-written from Eqs. 3.62 and 3.65 as

$$\begin{aligned}\hat{E}(t) = \cos(\omega_0 t)&\left[\sqrt{\frac{2\pi\hbar\omega_0}{Ac}}\int_0^\infty \left(\hat{a}_1 e^{-i\Omega t} + \hat{a}_1^\dagger e^{i\Omega t}\right)\frac{d\Omega}{2\pi}\right] \\ + \sin(\omega_0 t)&\left[\sqrt{\frac{2\pi\hbar\omega_0}{Ac}}\int_0^\infty \left(\hat{a}_2 e^{-i\Omega t} + \hat{a}_2^\dagger e^{i\Omega t}\right)\frac{d\Omega}{2\pi}\right].\end{aligned} \tag{3.67}$$

The terms multiplying $\cos(\omega_0 t)$ and $\sin(\omega_0 t)$ are the amplitudes of the field quadratures, respectively.

3.6 Quantum Description of the Interferometer

It is possible to describe the interferometer with the formalism just introduced, in order describe quantum noise, the main subject of this part of the work. In the following a Michelson interferometer with Fabry-cavities in the arms, as in Sect. 2.1.2, with input field E_{in} at angular frequency ω_0, is considered. The gravitational wave frequency, which lies in the range of 10–1000 Hz, is denoted by $f_{GW} = \frac{\Omega}{2\pi}$. The interaction of the wave with the optical system introduces sidebands at the angular frequencies $\omega_0 \pm \Omega$ into the output field at the dark port. As explained in [61], vacuum fluctuations enter from the anti-symmetric port of the detector and modulate the interferometer field in particular in the range of the gravitational wave signal, therefore introducing quantum noise. Quantum noise has been introduced in Sects. 2.2.4 and 2.2.5 with a semi-classical approach but the most practical way to describe it, when considering squeezing, is with the quantum formalism, as it will follow in this section. The vacuum fluctuations at the dark port of the detector can be expressed with the two-photon formalism as a field $\hat{E}_{vac}(t)$ entering from the output port equal to the field of Eq. 3.67.

This field recombines through the beam splitter with the main field inside the arms and then it is reflected back by the interferometer to the output port. The output field carrying the information on the recombination of the vacuum field with the interferometer [58] is given by

$$\begin{aligned}\hat{E}_{out}(t) = \cos(\omega_0 t)&\left[\sqrt{\frac{2\pi\hbar\omega_0}{Ac}}\int_0^\infty\left(\hat{b}_1 e^{-i\Omega t}+\hat{b}_1^\dagger e^{i\Omega t}\right)\frac{d\Omega}{2\pi}\right]\\ +\sin(\omega_0 t)&\left[\sqrt{\frac{2\pi\hbar\omega_0}{Ac}}\int_0^\infty\left(\hat{b}_2 e^{-i\Omega t}+\hat{b}_2^\dagger e^{i\Omega t}\right)\frac{d\Omega}{2\pi}\right].\end{aligned} \tag{3.68}$$

The input-output relations at the sideband angular frequency Ω are [58]

$$\begin{aligned}\hat{b}_1 &= \hat{a}_1 e^{2i\beta}\\ \hat{b}_2 &= (\hat{a}_2 - \mathcal{K}(\Omega)\hat{a}_1)e^{2i\beta} + \sqrt{2\mathcal{K}(\Omega)}\frac{h(\Omega)}{h_{SQL}(\Omega)}e^{i\beta}\end{aligned} \tag{3.69}$$

where $\hat{a}_1$ and $\hat{a}_2$ are the amplitude and phase operator defined in Sect. 3.5, and where the following quantities are involved:

- $\beta = \arctan\frac{\Omega}{\gamma}$: the net phase gained by the sideband at Ω inside the arm-cavities;
- $\gamma = \frac{Tc}{4L}$: half bandwidth of the arm cavity of length L and input mirror power transmittivity T;
- $h(\Omega)$: Fourier transform of the gravitational wave field;
- $h_{SQL}(\Omega)$: standard quantum limit (SQL) for GW detectors defined in Eq. 2.49;
- $\mathcal{K}(\Omega)$: the effective coupling constant, which relates the motion of the suspended mirrors to the output signal as

$$\mathcal{K}(\Omega) = \frac{2(P_{in}/P_{SQL})\gamma^4}{\Omega^2(\gamma^2+\Omega^2)} = \left(\frac{\Omega_{SQL}}{\Omega}\right)^2\frac{\gamma^2}{\Omega^2+\gamma^2}; \tag{3.70}$$

- Ω_{SQL} : standard quantum limit frequency, i.e. the frequency where the radiation pressure noise starts to dominates over the shot noise;
- P_{in} : input light power;
- P_{SQL} : power needed at the input of the interferometer to reach the SQL at the sideband frequency $\Omega = \gamma$

$$P_{SQL} = \sqrt{\frac{mL^2\gamma^4}{4\omega_0}}. \tag{3.71}$$

From Eq. 3.69, it can be noticed that the gravitational wave signal is carried by the quadrature $\hat{b}_2$, i.e. by the *phase* quadrature. Moreover, it can be noticed that this quadrature depends on both the quadratures of the vacuum state entering from the dark port. The quantum noise spectral density, derived from Sects. 2.2.4 and 2.2.5, can be expressed using Eq. 3.69 as

$$S_{qn}(\Omega) = \left(\frac{1}{\mathcal{K}(\Omega)} + \mathcal{K}(\Omega)\right)\frac{h_{SQL}^2}{2} = S_{shot}(\Omega) + S_{rp}(\Omega). \tag{3.72}$$

The two components of quantum noise can be recognized from Eq. 3.72: the shot noise at high frequencies $S_{shot} = \frac{1}{\mathcal{K}(\Omega)}\frac{h_{SQL}^2}{2}$ and the radiation pressure noise at low frequencies $S_{rp} = \mathcal{K}(\Omega)\frac{h_{SQL}^2}{2}$. In this Section, a Fabry-Perot recycled Michelson interferometer has been considered while the quantum description of a signal-recycled interferometer is given in [62].

It is convenient to express the input-output relations of Eq. 3.69 with a matrix formalism as $\overline{b} = T_{itf}\overline{a} + b_{GW}$

$$\begin{pmatrix}\hat{b}_1\\ \hat{b}_2\end{pmatrix} = T_{itf}\begin{pmatrix}\hat{a}_1\\ \hat{a}_2\end{pmatrix} = \begin{pmatrix}e^{2i\beta} & 0\\ -\mathcal{K}(\Omega)\cdot e^{2i\beta} & e^{2i\beta}\end{pmatrix}\begin{pmatrix}\hat{a}_1\\ \hat{a}_2\end{pmatrix} + \begin{pmatrix}0\\ \sqrt{2\mathcal{K}}\frac{h(\Omega)}{h_{SQL}(\Omega)}e^{i\beta}\end{pmatrix}. \tag{3.73}$$

From here it is evident the effect of the interferometer on a vacuum quantum field.

3.7 Squeezed States to Reduce Quantum Noise

It is now interesting to investigate what happens for the particular quantum states described in Sect. 3.1.3, the squeezed vacuum states. Actually, thanks to these states, as predicted by Caves in 1981 [61], it is possible to reduce the quantum noise defined in Eq. 3.72 and to demonstrate this it is very practical to use the two-photon formalism expressed in matrix form, as in Eq. 3.73. Actually, vacuum squeezed states can be described by the matrix $\overline{a}'$ as

$$\begin{aligned}\begin{pmatrix}\hat{a}_1\\ \hat{a}_2\end{pmatrix}' &= R(\theta)S(r)R(\theta)^\dagger\begin{pmatrix}\hat{a}_1\\ \hat{a}_2\end{pmatrix} = \\ &= \begin{pmatrix}\cos^2(\theta)e^r + \sin^2(\theta)e^{-r} & \cos(\theta)\sin(\theta)(e^r - e^{-r})\\ \cos(\theta)\sin(\theta)(e^r - e^{-r}) & \sin^2(\theta)e^r + \cos^2(\theta)e^{-r}\end{pmatrix}\begin{pmatrix}\hat{a}_1\\ \hat{a}_2\end{pmatrix}\end{aligned} \tag{3.74}$$

where $R(\theta)$ is the rotation matrix and $S(r)$ is the squeeze operator for $\theta = 0$

$$R(\theta) = \begin{pmatrix} \cos\theta & -\sin\theta \\ \sin\theta & \cos\theta \end{pmatrix} \qquad S(r) = \begin{pmatrix} e^{r} & 0 \\ 0 & e^{-r} \end{pmatrix}. \tag{3.75}$$

By substituting Eq. 3.74 in Eq. 3.73, the effect of the interferometer on squeezed vacuum states is obtained.

3.7.1 Frequency Independent Squeezing

If the angle of the ellipse of squeezed vacuum states introduced by the detector output port is fixed, one has the so-called *frequency independent squeezing*. By choosing $\theta = \frac{\pi}{2}$, one is introducing vacuum states with reduced phase fluctuations $\hat{a}_2$, called $\hat{X}_2$ in Sect. 3.1.3. Therefore, from Eqs. 3.74 and 3.73, the following quantum noise spectral density is obtained [58, 61]

$$S_{qn}^{\theta=\frac{\pi}{2}}(\Omega) = \left(\frac{1}{\mathcal{K}(\Omega)}e^{-2r} + \mathcal{K}(\Omega)e^{2r}\right)\frac{h_{SQL}^2}{2} = S_{shot}(\Omega)e^{-2r} + S_{rp}(\Omega)e^{2r}. \tag{3.76}$$

Comparing this equation to Eq. 3.72 that the shot noise is reduced but the radiation pressure noise is increased. On the other hand, by choosing $\theta = 0$, one is introducing vacuum states with reduced amplitude fluctuations $\hat{a}_1$, called $\hat{X}_1$ in Sect. 3.1.3. Analogously, the quantum noise spectral density is given by [58, 61]

$$S_{qn}^{\theta=0}(\Omega) = \left(\frac{1}{\mathcal{K}(\Omega)}e^{2r} + \mathcal{K}(\Omega)e^{-2r}\right)\frac{h_{SQL}^2}{2} = S_{shot}(\Omega)e^{2r} + S_{rp}(\Omega)e^{-2r}. \tag{3.77}$$

In this case the radiation pressure noise is reduced but the shot noise is increased. The interesting result is that introducing frequency-independent squeezed states into the interferometer can reduce quantum noise at high or low frequencies, depending on the direction in which the states are squeezed. The respective amplitude spectral density $h_{qn}^{\theta}(\Omega) = \sqrt{S_{qn}^{\theta}(\Omega)}$ is shown on the left side of Fig. 3.11.

3.7.2 Frequency Dependent Squeezing

From Eqs. 3.76 and 3.77, one can infer that having squeezed states whose θ can rotate with a frequency-dependence to the sideband frequency Ω, can allow a quantum noise reduction over the entire detection spectrum. The condition for achieving broadband quantum noise reduction for a Fabry-Perot recycled ITF is given by Kimble et al. [58]

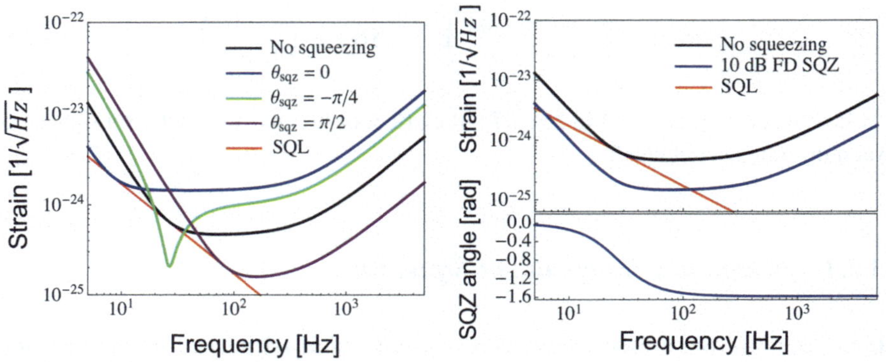

Fig. 3.11 Amplitude spectral density of the quantum noise curve $h_{qn} = \sqrt{S_{qn}}$. The black curves represent the quantum noise of a dual-recycled interferometer and the red curves the standard quantum limit (SQL) in both the plots. On the left, the effect of frequency-independent squeezing injection on the quantum noise is shown for different fixed angles: $\theta = 0$ in blue, $\theta = \frac{\pi}{2}$ in purple and $\theta = -\frac{\pi}{4}$ in green. On the right, the effect of 10dB-frequency-dependent squeezing injection is shown (blue curve). The rotation of the squeezing angle as a function of the frequency is shown on the right bottom plot. Figure from [63]

$$\theta(\Omega) = -\operatorname{arccot}\mathcal{K}(\Omega). \tag{3.78}$$

In this case, the power spectral density of the quantum noise results to be

$$S_{qn}^{broad}(\Omega) = \left(\frac{1}{\mathcal{K}(\Omega)}e^{-2r} + \mathcal{K}(\Omega)e^{-2r}\right)\frac{h_{SQL}^2}{2} = S_{shot}(\Omega)e^{-2r} + S_{rp}(\Omega)e^{-2r}. \tag{3.79}$$

The respective amplitude spectral density $h_{qn}^{broad}(\Omega) = \sqrt{S_{qn}^{broad}(\Omega)}$ is shown on the right side of Fig. 3.11.

The method currently used to introduce to squeezed vacuum states, a frequency dependence, was introduced for the first time in 2001 by Kimble et al. [58] and consists in reflecting FIS state by a detuned high finesse Fabry-Perot cavity. In this way, only quantum fluctuations that have a frequency within the bandwidth of the cavity enter the cavity and are therefore dephased. Instead, the other fluctuations are reflected directly from the cavity. This cavity is therefore called a filter cavity because it filters the incident field. For the rotation to occur around the right frequency, it is necessary for the cavity to be detuned with respect to its resonance by $\Delta\omega_{fc} = \omega_{fc} - \omega_0$, where ω_0 is the carrier frequency and ω_{fc} is the frequency for which the cavity is resonant. This causes sidebands at $\omega_0 \pm \Omega$ to see a different cavity reflectivity because of this asymmetry, as shown in Fig. 3.12. In the following the cavity will be considered lossless. The reflectivity of a Fabry-Perot cavity is defined in Eq. 2.16 following the one-photon formalism and can be expressed as [64]

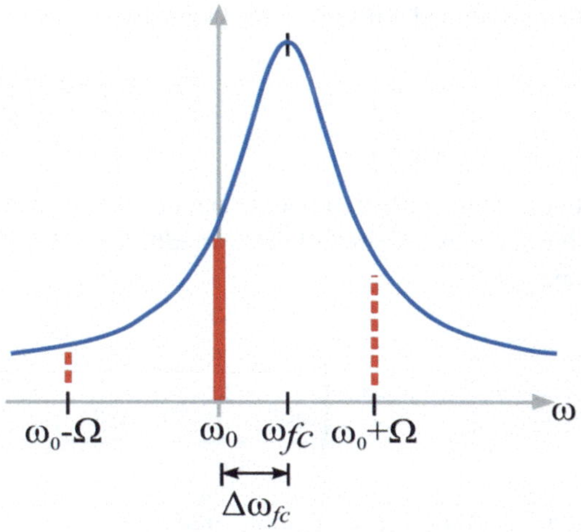

Fig. 3.12 Filter cavity transmitted power: ω_{fc} is the resonance frequency and ω_0 is the carrier frequency. The difference between the cavity and the carrier frequencies is the filter cavity detuning $\Delta\omega_{fc} = \omega_{fc} - \omega_0$. The sidebands at $\omega_0 \pm \Omega$ are experiencing an asymmetric reflection due to the detuning. Figure from [53]

$$r_{fc}(\Omega) = \frac{r_1 - r_2 t_1^2 e^{-i\Phi(\Omega)}}{1 - r_1 r_2 e^{-i\Phi(\Omega)}} \tag{3.80}$$

where $r_{1,2}$ are the reflectivities of the cavity input and end mirrors respectively and $\Phi(\Omega) = 2kL_{fc}(\Omega - \Delta\omega_{fc})$. To describe the effect the filter cavity has on vacuum squeezed states, it is appropriate to use the two-photon formalism [53, 64]. The filter cavity transfer function is given by Evans et al. [64] and Kwee et al. [53]

$$T_{fc}(\Omega) = \frac{1}{2}\begin{pmatrix} r_{fc}(\Omega) + r_{fc}(-\Omega)^* & i\left(r_{fc}(\Omega) - r_{fc}(-\Omega)^*\right) \\ -i\left(r_{fc}(\Omega) - r_{fc}(-\Omega)^*\right) & r_{fc}(\Omega) + r_{fc}(-\Omega)^* \end{pmatrix}. \tag{3.81}$$

Since the reflectivities of the cavity are complex numbers, they can be expressed as

$$r_{fc}(\Omega) = \rho(\Omega)e^{i\alpha(\Omega)} \qquad r_{fc}(-\Omega) = \rho(-\Omega)e^{i\alpha(-\Omega)} \tag{3.82}$$

and it is also convenient to define the sum and difference for the amplitude and phase of the reflectivities as

$$\rho_{p,m} = \frac{\rho(\Omega) \pm \rho(-\Omega)}{2} \qquad \alpha_{p,m} = \frac{\alpha(\Omega) \pm \alpha(-\Omega)}{2}. \tag{3.83}$$

Considering that the cavity has high finesse and is close to resonance, the following relationships apply

$$e^{i\Phi} \sim 1 - i\Phi \qquad r_1 r_2 \sim 1 - \frac{t_1^2}{2} \tag{3.84}$$

that combined with Eq. 2.16, yields to

$$r_\pm = \frac{-1 + i\Phi(\pm\Omega)/t_1^2}{1 + i\Phi(\pm\Omega)/t_1^2}. \tag{3.85}$$

For a cavity with high finesse and near resonance, it is proven that $r_- = r_+$, meaning that $\rho_p = \rho_m$. Using all these conditions in Eq. 3.80 and by substituting in Eq. 3.81, one gets

$$T_{fc} = \rho_p e^{i\alpha_m} \begin{pmatrix} \dfrac{e^{i\alpha_p} + e^{-i\alpha_p}}{2} & -\dfrac{e^{i\alpha_p} - e^{-i\alpha_p}}{2i} \\ \dfrac{e^{i\alpha_p} - e^{-i\alpha_p}}{2i} & \dfrac{e^{i\alpha_p} + e^{-i\alpha_p}}{2} \end{pmatrix} = \rho_p e^{i\alpha_m} R_{\alpha_p} \tag{3.86}$$

with $\alpha_p(\Omega)$ given by Kimble et al. [58]

$$\alpha_p(\Omega) = \arctan\left(\frac{2\gamma_{fc}\Delta\omega_{fc}}{\gamma_{fc}^2 - \Delta\omega_{fc}^2 + \Omega^2}\right). \tag{3.87}$$

This shows that the effect of the filter cavity, in the two-photon formalism, is to introduce a rotation with angle α_p to the states reflected from it, in analogy with Eq. 3.75.

For a lossless cavity where $\Omega << \gamma_{fc}$ the detuning and the cavity linewidth have to be the same and equal [58] in order to have anti-squeezing for low frequencies and squeezing for high frequencies:

$$\Delta\omega_{fc} = \gamma_{fc} = \frac{\Omega_{SQL}}{\sqrt{2}}. \tag{3.88}$$

In conclusion a detuned filter cavity transforms FIS states into FDS states.

3.8 Conclusion

This chapter derives a quantum model of the interferometer. This model allows to describe the quantum noise as quantum fluctuations of the vacuum states entering the output port of the detector. The shot noise is represented by phase quadrature vacuum fluctuations and the radiation pressure noise by amplitude quadrature vacuum fluctuations. However, vacuum squeezed states are needed in order to reduce the detector quantum noise. The generation and the detection principle of these quantum states has been presented. In particular, an homodyne detector is used in Virgo to characterize these states, as it will be detailed in the next chapter. The degradation mechanisms of these states, such as optical losses and phase noise have also been detailed. Several sources of optical losses relative to the Virgo systems will be discussed in the following chapters. We have shown that by using a filter cavity it is possible to introduce a frequency dependence to the squeezed states and therefore obtain a broadband reduction of quantum noise in the detectors. This is what has been implemented in Virgo. The experimental setup of the whole frequency-dependent squeezing system is reported in the next chapter.

Chapter 4
Quantum Noise Reduction System Overview in Advanced Virgo Plus

In Advanced Virgo frequency independent squeezing was injected during O3 run, reducing the quantum noise due to shot noise at high frequencies and consequently increasing noise at low frequencies [65, 66]. The path to achieve ellipse rotation for FDS in the frequency band useful for gravitational wave detectors, i.e. 10–100 Hz, is composed of gradual achievements. Frequency dependent squeezing was first demonstrated in 2005 by Chelkowski et al, via a 0.5 m filter cavity and 1.47 MHz linewidth [67]. Subsequently, in 2015 Oelker et al achieved a rotation at 1.2 kHz with a 2 m filter cavity with finesse of 30,000 [68]. Finally, the rotation of squeezing states in the frequency range of gravitational wave detectors interest was demonstrated in 2020 at Massachusetts Institute of Technology (MIT) and at National Astronomical Observatory of Japan (NAOJ). At MIT, a rotation at 30 Hz was demonstrated using a 16 m filter cavity [69]. At NAOJ, we demonstrated a rotation of the squeezing ellipse at 90 Hz with a 300 m suspended cavity [70]. The latter result represents a 1:1 scale prototype of the setup being implemented in the actual detectors for quantum noise reduction for the observing run O4.

In this chapter, the transition from FIS to FDS in the Virgo detector is described. At first, the limitations of FIS system used during the run O3 are presented. Subsequently, the improvements on the system and the description of the experimental setup of FDS on the Virgo site are described.

4.1 Frequency Independent Squeezing During O3

During the run O3, frequency independent squeezing was successfully injected in Advanced Virgo to reduce quantum noise in the high frequency detection bandwidth. The conceptual scheme of the system implemented in the detector is shown in Fig. 4.1.

In AdV, the input beam had a power of 18 W. The gravitational wave signal is detected at the output port of the interferometer using a DC readout technique. An

E. Polini, *Broadband Quantum Noise Reduction in Advanced Virgo Plus*,
Springer Theses, https://doi.org/10.1007/978-3-031-95143-5_4

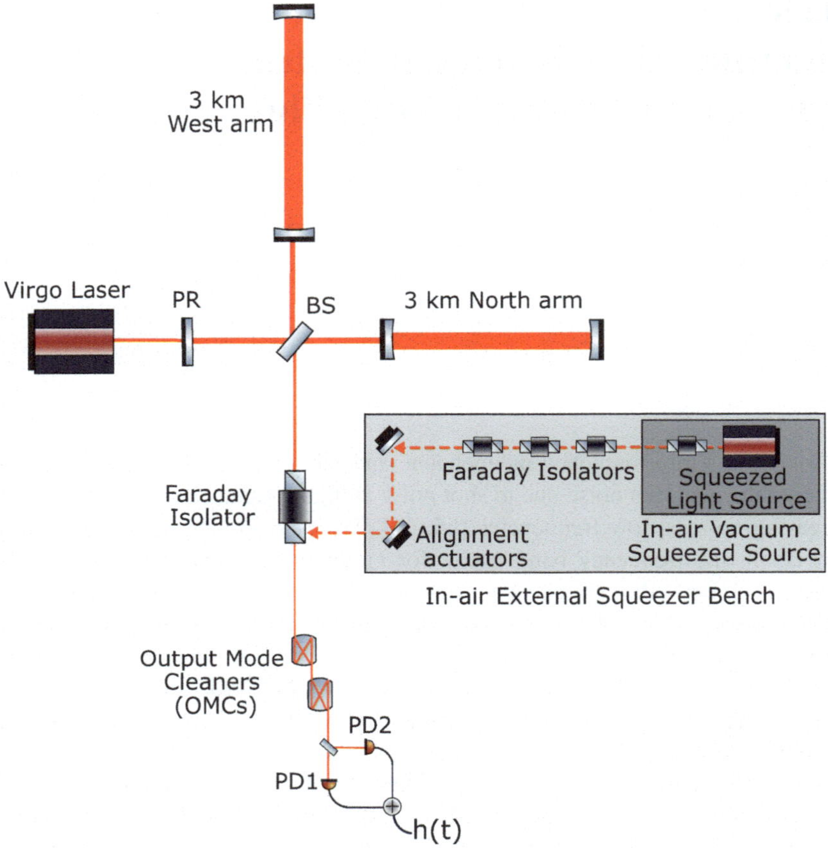

Fig. 4.1 Conceptual scheme of FIS implemented during O3 in AdV

in-air optical bench hosts the squeezed light source. The laser that generates the squeezing is locked to the Virgo's laser frequency with a phase lock loop (PLL) so that the squeezing can interact with the interferometer beam to reduce quantum noise. The squeezed light first passes through a Faraday Isolator (FI) inside the Vacuum Squeezed Source. Then, the beam passes through a triplet of low-loss Faraday Isolators with an isolation factor of 40 dB and 1% single-path losses each. FIs are needed to attenuate the effect of backscattered light from the interferometer (see Part III). FIS is then injected into the interferometer through a Faraday Isolator located on the suspended detection bench SDB1. The squeezed beam needs to be well aligned and matched to the ITF beam. The alignment is performed with remotely controllable piezo-electric actuators mounted on two mirrors located on the External Squeezer Bench.

With this technique, a shot noise reduction of (3.2 ± 0.1) dB has been observed in the region between $\sim$300 Hz and 3.2 kHz, as shown in Fig. 4.2. The injection of FIS in Advanced Virgo has been operated with a duty cycle of 99% for the recorded

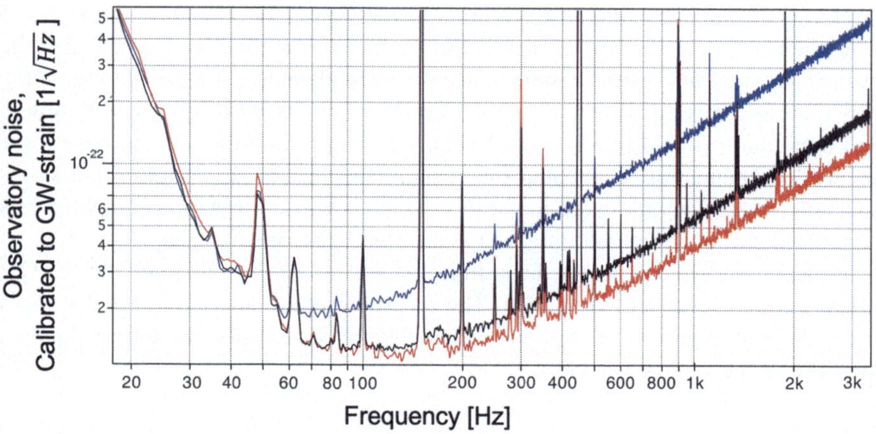

Fig. 4.2 Measured spectral strain sensitivity of Advanced Virgo detector. The black curve corresponds to the detector sensitivity without squeezing injection. The blue curve corresponds to the sensitivity with anti-squeezing injection: the shot noise is increased by (8.5 ± 0.1) dB. The red curve corresponds to the sensitivity with squeezing injection: the shot noise is reduced by (3.2 ± 0.1) dB. The level of injected squeezing was about 10 dB [65]

science data during the observing run O3. This lead to a 5–8% improvement of the binary neutron star horizon, which corresponds to 16–26% increase of detection rate [65].

4.1.1 Frequency Independent Squeezing Limitations

However, this technique has some limitations.

At first, as can be seen from the spectrum in Fig. 4.2, the radiation pressure noise increases slightly for frequencies below 80 Hz: the red curve representing the interferometer sensitivity with squeezing lies above the black curve that shows the sensitivity without squeezing. This was not an issue during O3, but it is a problem for an interferometer radiation pressure limited, as it will be the case for AdV+ in O4. In this case, it is needed to inject squeezed states of light that rotate as a function of frequency (FDS). This is done for AdV+.

Moreover, scattered light from different sources is another limitation that has been observed. The effect on the sensitivity due to scattered light scales as the square root of the power of the scattered field and with the effective displacement noise of the scattering surface (as will be discussed in Part III). Faraday Isolators are used to attenuate the back-scattered light. On one hand, we need to minimize their number as they introduce optical losses, on the other hand, we also need to ensure that they attenuate the scattered light enough not to introduce additional noise. Furthermore, during O3 it was observed that the scattered light from the last FI on the squeezing bench near the detection towers back-scattered the spurious beam from the interferometer. This is mainly due to the fact that the FI was attached to the ground and therefore vibrates more than suspended elements. In view of O4, a study was therefore carried out on the optimal number of FIs on the squeezing system and how many should be suspended [71]. The ideal compromise is to have three suspended attenuation stages and one ground-attached FI. In addition, numerous spurious beams were observed in the FIS, which is why a mitigation campaign was conducted during commissioning for O4, as it will described in Chap. 7.

Finally, the optical losses degrading squeezing were quite high during O3: $(36.5 \pm 4.5)\%$ were measured [65]. The measured phase noise is (45 ± 40) mrad rms [65]. Improvements of the optical elements to reduce losses have been done on the FDS system for the run O4.

4.2 Frequency Dependent Squeezing for O4

The conceptual scheme of the FDS system in AdV+ is shown in Fig. 4.3.

The frequency-independent squeezed vacuum states are generated by the same source used during O3, which is located on an in-air external bench (EQB1) attached to the ground. As can be seen in the scheme, the beam passes through a low-loss Faraday Isolator on EQB1 bench. The new FI has losses of 1%, which is lower than the 2% losses of the previous one. Subsequently, the beam passes through vacuum-suspended benches (SQB1, SQB2) where it is reflected by a Faraday Isolator. Then the squeezed states are reflected by a 285-meter vacuum-suspended detuned cavity, where they acquire the frequency dependence (see Chap. 5). Then they are sent towards the interferometer by passing through two Faraday isolators on the suspended benches.

4.2.1 Losses Budget and Improvements

The fact that the Faraday Isolators are suspended and one has been replaced reduce optical losses and scattered light in the system compared to FIS system in O3 run. In addition, replacing the optics in the system with better quality optics further reduces optical losses. Actually, the optical elements of the suspended system, such as mirrors, lenses and waveplates are optimized in number to reduce optical losses at each surface. Moreover, the mirrors are super-polished (flatness $< \lambda/50$ RMS, roughness < 0.1 nm RMS) and coated at the Laboratoire des Matériaux Avancés (LMA) with high reflecting coatings, which present a residual transmission of few ppm. In addition, the output mode cleaner cavities have been replaced with a single low losses high finesse cavity (see Part IV). Phase noise, on the other hand, may be higher due to the contribution of the control loops needed to control the more complex system.

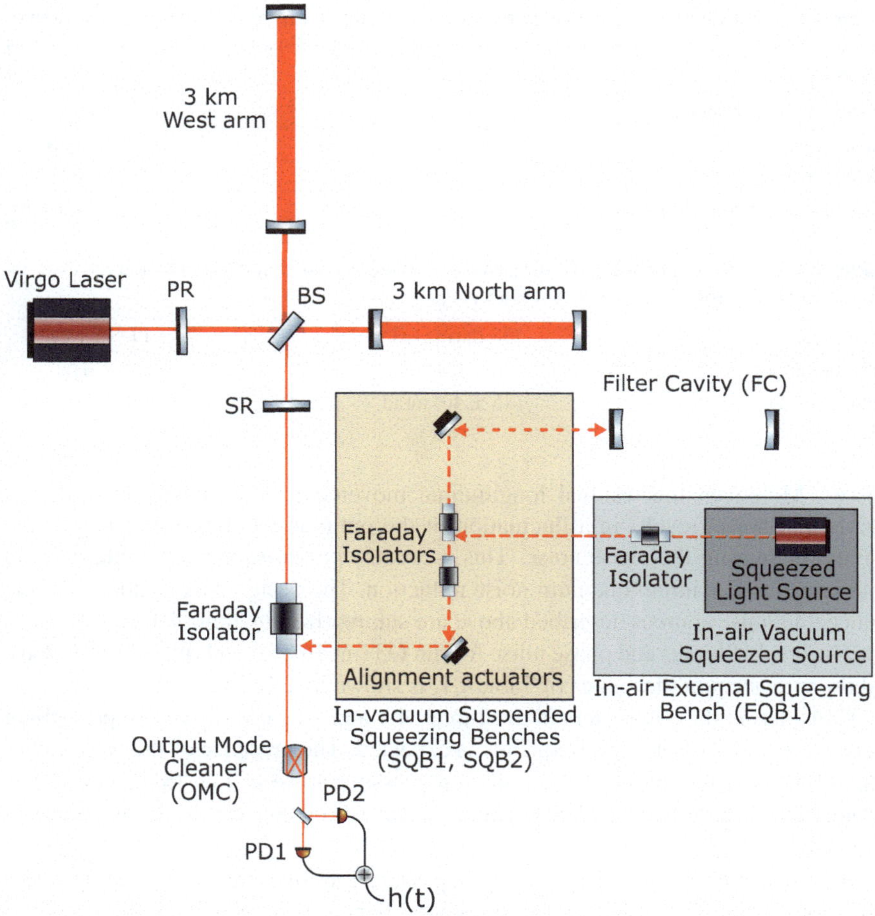

Fig. 4.3 Conceptual scheme of FDS implemented for O4 in AdV+. The external squeezing bench (EQB1) hosts the vacuum squeezed source. Then there are two in-vacuum suspended benches that send the squeezing into the 285 m filter cavity, parallel to the North arm of the interferometer. FDS is therefore injected in the detector

One must also consider the losses introduced by the filter cavity. The optical losses introduced by the cavity degrade squeezing in both ways: either by mixing the two quadratures (phase noise) or by mixing the standard vacuum with the squeezed vacuum (optical losses). It has been shown that the filter cavity losses are inversely proportional to the cavity length [72]. The round-trip losses for our 285 m-long cavity, considering the flatness and the roughness of its mirrors, are 40 ppm, which is very low and compatible with the losses budget [73, 74].

Furthermore, the beam must be matched not only to the interferometer, but also to the filter cavity. The fraction of squeezing field not matched with the cavity bypasses the cavity without experiencing frequency rotation introduces optical

Table 4.1 Filter cavity parameters required for O4 [74], for 10 dB of injected squeezing. Round-trip losses and mode matching are optical losses and the longitudinal residual motion accounts as phase noise

Filter cavity parameter	Value
Round-trip losses	40 ppm
Mode matching	98%
Longitudinal residual motion	0.3 pm rms

Table 4.2 Comparison between the measured squeezing degradation during O3 and the expectations for O4 [74], for 10 dB of injected squeezing

Parameter	FIS during O3	FDS for O4
Optical losses	$(36.5 \pm 4.5)\%$	$(22 \pm 4)\%$
Phase noise	(45 ± 40) mrad	(40 ± 20) mrad

losses. Moreover, the residual longitudinal movement of the cavity mirrors, i.e. the length noise, results in a fluctuation of the optimal detuning frequency of the cavity, increasing the phase noise. This prevents optimal squeezing angle rotation and prevents maximum quantum noise reduction. The design specifications for the squeezing noise sources described above are summarized in Table 4.1.

The level of losses and phase noise for run O3 (measured) and run O4 (expected), taking into account the values of Table 4.1, is shown in Table 4.2.

Considering the parameters in the Tables 4.1 and 4.2, a sensitivity improvement between 6 and 7 dB at high frequency is expected. The presence of the filter cavity should allow a reduction of the radiation pressure noise at low frequencies. The projection of the effect of FDS injection on interferometer sensitivity is shown in Fig. 4.4.

It is also possible to inject FIS by bypassing the filter cavity: a flipping retro-reflector mirror is installed on the suspended bench. It reflects back the squeezed beam and sends it directly into the interferometer.

4.2.2 Frequency Dependent Squeezing Infrastructure on Site

The implementation of the FDS system in Virgo required considerable infrastructure work. A scheme of the system at the site is shown in Fig. 4.5. The squeezing is produced on the in-air EQB1 bench. It then passes onto the suspended SQB1 bench via a viewport. EQB1 and SQB1 are located in the detection lab, called like this because it also hosts two other benches of the detection system: SDB2 and EDB. The SQB1 bench is then connected to SQB2 via a 12 m pipe. Both suspended benches are inside in-vacuum mini-towers. SQB2 is located in a purpose-built clean room. SQB2 is linked by a 40-meters pipe to the cavity's input mirror (FCIM). After the 285-meters filter cavity pipe there is the filter cavity's end mirror (FCEM). The

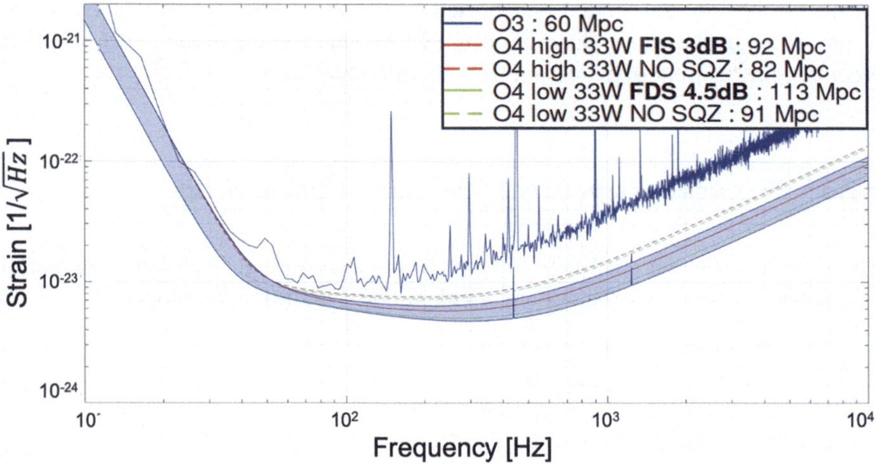

Fig. 4.4 Comparison between O3 Virgo sensitivity curve and O4 sensitivity predictions with no squeezing, FIS and FDS. The light purple area represents the range between pessimistic and optimistic sensitivity estimations for O4. In blue the data of the best sensitivity achieved with FIS injected in the ITF during the run O3 (60 Mpc). In red the projection of O4 sensitivity pessimistic, considering 3 dB of FIS injected for the continuous line and no squeezing for the dashed line. In green the projection of O4 sensitivity optimistic, considering 4.5 dB of FDS injected for the continuous line and no squeezing for the dashed line

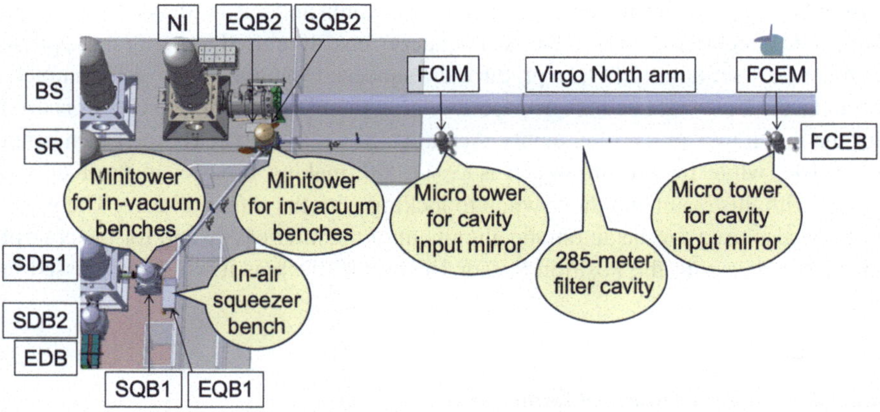

Fig. 4.5 FDS system infrastructure in AdV+ [74]. SDB1, SDB2 and EDB are detection benches. SR, BS and NI are suspended interferometer optics. EQB1, EQB2 and FCEB are in-air ground-attached benches. SQB1 and SQB2 are in vacuum suspended benches. FCIM and FCEM are the filter cavity in-vacuum suspended mirrors

cavity mirrors are suspended and hosted inside in-vacuum micro-towers, inside clean areas. In the same clean area of FCEM, there is an in-air bench (FCEB) receiving the beam in transmission from the filter cavity.

4.2.3 Frequency Dependent Squeezing Optical Setup

The simplified optical scheme of the system is shown in Fig. 4.6, where the several beams needed to control the system are identified with different colors:

- the green auxiliary beam (green line);
- the sub-carrier beam (orange line);
- the vacuum squeezed beam (dashed red line);
- the local oscillator (blue line);
- the spurious beam coming from the interferometer (purple line).

A fraction of the green and sub-carrier beams (called pick-off) is extracted in order to control the benches and the filter cavity in their nominal positions. The pick-off beams are sensed on photodiodes (PDs), position sensor devices (PSDs), quadrants (QDs) and cameras (CAM), that are shown on Fig. 4.6.

4.2.3.1 Green Auxiliary Beam (GR)

The auxiliary green beam has a wavelength of 532 nm and it is produced by a second harmonic generator (SHG) in the AEI squeezer box. It is used to pump the Optical Parametric Oscillator (OPO) to produce the squeezed vacuum states. A green pick-off exits the squeezer box and propagates within the FDS system. This beam is used for the initial lock of the filter cavity. Actually, the cavity finesse for the green beam is $\sim$100, while for the infrared it is $\sim$11,000, making it easier to pre-lock the cavity with this beam and then switch to locking with the infrared beam. The green beam is also used for the automatic alignment of the filter cavity and for the control of the benches with the Position Sensor Devices (PSDs), as it will be discussed in Sect. 5.2.2.

4.2.3.2 Vacuum Squeezed Beam (SQZ)

The squeezed beam has a wavelength of 1064 nm, the same as Virgo in order to effectively reduce quantum noise. To this purpose, its phase is locked to that of Virgo's main laser by means of the Phase Lock Loop (PLL).

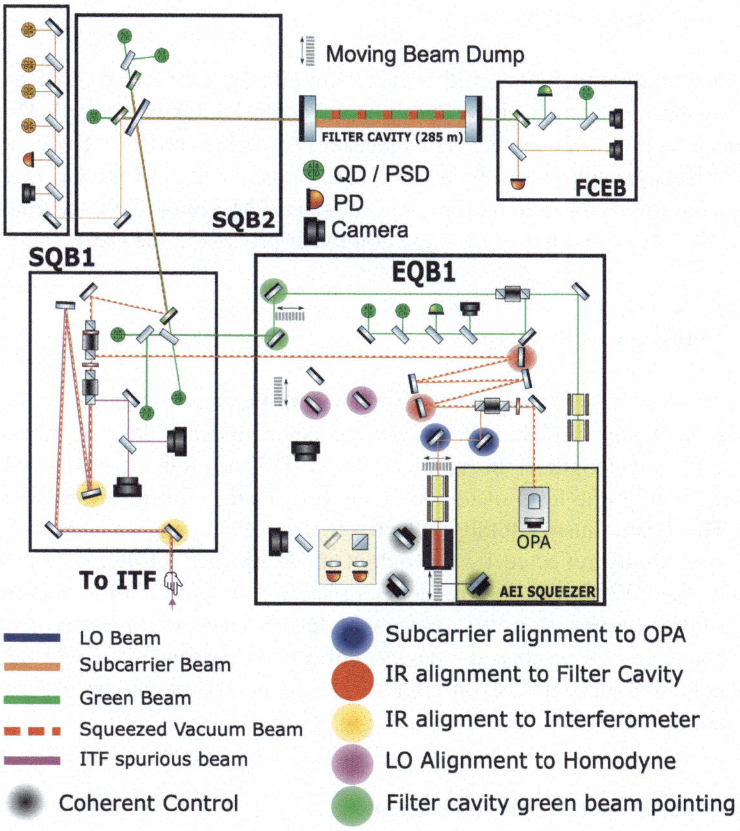

Fig. 4.6 Simplified optical scheme of the FDS system in Advanced Virgo Plus operated on its first configuration. Credit M. Vardaro

4.2.3.3 Bright Alignment Beam (BAB)

A bright beam mimicking the squeezing beam is used in order to align the optics and it is called Bright Alignment Beam (BAB). It is produced when the OPO crystal is not pumped with the green beam but with an infrared beam. This beam is amplified in the cavity and the transmitted beam is called BAB and it has the same optical properties and alignment as the squeezed field and a power of a few mW.

4.2.3.4 Local Oscillator Beam (LO)

The local oscillator is a pick-off of the main beam of the squeezer laser (at the same frequency as the Virgo laser) that propagates on EQB1 to the homodyne where it is used for squeezing measurements.

4.2.3.5 Coherent Control Beam (CC)

In order to stabilize the angle of the squeezing ellipse, an infrared auxiliary beam mimicking the squeezing beam, called coherent control (CC), is required. This beam is produced by a laser inside the AEI squeezer box and is shifted in frequency with respect to the squeezing beam by a PLL. The frequency shift of the CC beam with respect to the squeezing field has been chosen to be 4 MHz in order to be transmitted by the OMC cavity and not to spoil the non linear process in the OPO.

4.2.3.6 Sub-carrier Beam (SC)

In order to lock the FC more precisely than with the green beam, it is necessary to use an IR beam mimicking the squeezed beam. Indeed, the sub-carrier beam has the same wavelength as the squeezed beam. This beam is superimposed on the squeezing beam at the level of the OPO and propagates with it towards the whole system. This beam must be totally reflected by the OPO cavity, in resonance inside the FC, and totally reflected by the output mode cleaner (OMC) so that it does not pollute the GW signal. The frequency offset to satisfy these three requirements is $\sim$1.26 GHz, fixed with a PLL. However, during squeezing measurements, it is therefore not possible to lock the cavity with the SC because it would otherwise reach the homodyne detector on EQB1. The lock is kept on the green in that configuration.

4.2.3.7 Interferometer Spurious Beam

The output beam of the interferometer contains both s-polarized light, which contains the GW signal and p-polarized light, which accounts for the ITF defects. The Faraday Isolator on the detection system rejects the p-polarized light which is sent to the squeezing system. This beam is the ITF spurious beam and it has the same properties as the s-polarized beam. It is measured on the squeezing system in order to adjust the alignment and the mode matching of the squeezed beam with the interferometer before its injection (see Chap. 9). SQB1 hosts two cameras, used as sensors for the alignment loop, and two piezo-actuated tip-tilt platforms acting on two steering mirrors, to adjust the alignment.

4.3 External Squeezing Bench 1 (EQB1)

The External Squeezing Bench 1 (EQB1) is an L-shaped ground-attached bench located in the detection laboratory in the Virgo central building. Its dimension and shape have been chosen as a compromise between the available space and functional requirements. The ground seismic vibrations of the bench are attenuated thanks to elastomeric dumpers made by sorbothane cylinders, located under the table legs. This bench hosts several elements, as shown on Fig. 4.6:

- the vacuum squeezed source;
- phase locked loops optics and electronics;
- the sub-carrier field source;
- the homodyne detector;
- optics to steer and match the vacuum field, the sub-carrier field and the local oscillator for the homodyne detection on EQB1;
- optics and electronics for steering, mode matching and frequency shift the green field used for the first stage longitudinal and angular control of the filter cavity;
- optics for the *delay line* that simulates entirely on EQB1 the squeezing path up to the filter cavity for diagnostic purposes.

4.3.1 *Vacuum Squeezed Source*

The vacuum squeezed source is contained in a 1 m^2 box on EQB1 bench and its optical scheme during O3 is shown in Fig. 4.7 and all the details can be found in [75]. Inside the squeezer there are two lasers: the main laser that produces the

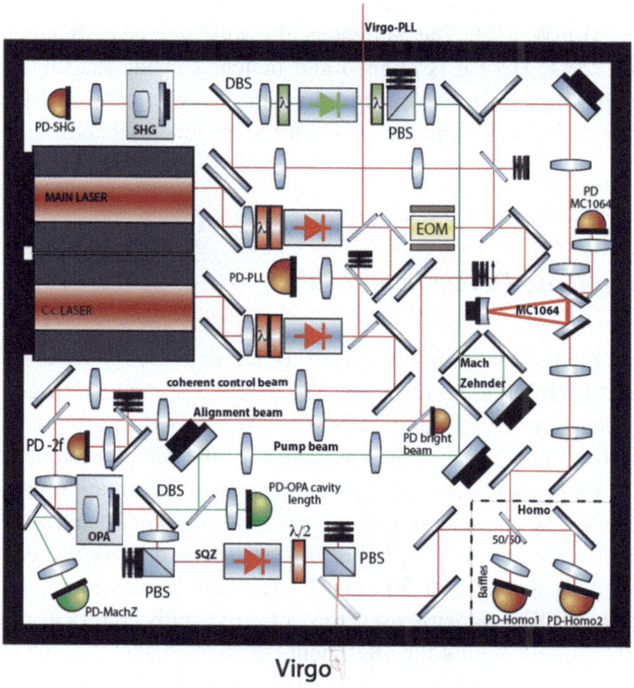

Fig. 4.7 Schematic of the experimental setup of the AEI squeezer for Advanced Virgo [76]. The setup for Advanced Virgo Plus is almost the same with only minor changes

vacuum squeezed states and the coherent control (CC) laser that is needed to control the ellipse angle. These two lasers are locked in phase with a phase locked loop (PLL) with a shift of 4 MHz. A fraction of the infrared beam at 1064 nm from the main laser is converted in green at 532 nm with a second-harmonic generation (SHG) cavity. The generated 532 nm field passes through a Faraday Isolator to protect the SHG from back reflections and a compact Mach–Zehnder interferometer (MZ) in the green path is used to control and stabilize the pump power impinging onto the OPO, as described in Sect. 3.2. The pump power determines the level of parametric (de-)amplification inside the OPO and it varies between 1.5 and 12 mW by changing the offset of the MZ operating point. The vacuum squeezed states are produced by parametric down-conversion in a periodically-poled potassium titanyl phosphate (PPKTP) crystal contained in the OPO cavity which is a doubly resonant cavity (both for IR and green). A pick-off of the main laser is used to stabilize the squeezed light source to the fundamental interferometer laser frequency by means of an external phase locking loop.

4.3.2 Sub-carrier Source

The sub-carrier (SC) field is produced by a Mephisto laser. Its power has been chosen to be 10 mW [74]. The SC enters the squeezer box after being reflected by the FI on EQB1. Here it is aligned and matched inside the OPO cavity to be completely superimposed to the squeezed beam. It is then reflected from the OPO and propagates to the cavity and then to the interferometer, where it will be reflected by the OMC.

4.3.3 Homodyne Detector on EQB1

As described in Sect. 3.3, an homodyne (HD) detector is needed to measure squeezed states in different quadratures. The external homodyne detector on EQB1 is needed to characterize FIS and FDS states before sending them to the ITF in order to optimize the losses. The optical scheme of the detector that is enclosed into a (250 $\times$ 225) mm box is shown in Fig. 4.8 and more information can be found in [77]. The squeezed beam recombines with the local oscillator beam on a 50/50 beam splitter. The reflected and transmitted beams are focused two photodiodes through two lenses with focal length of 50 mm. There is a flip-mirror that allows to send the beam on the EQB1 cameras used to overlap visually the local oscillator with the BAB or the SC, for tuning the alignment (see Sect. 5.4.1). The HD photodiodes are high quantum efficiency ($\eta > 99\%$) InGaAs 500 μm-diameter detectors at 1064 nm, with anti-reflective coating for an angle of incidence of 20°.

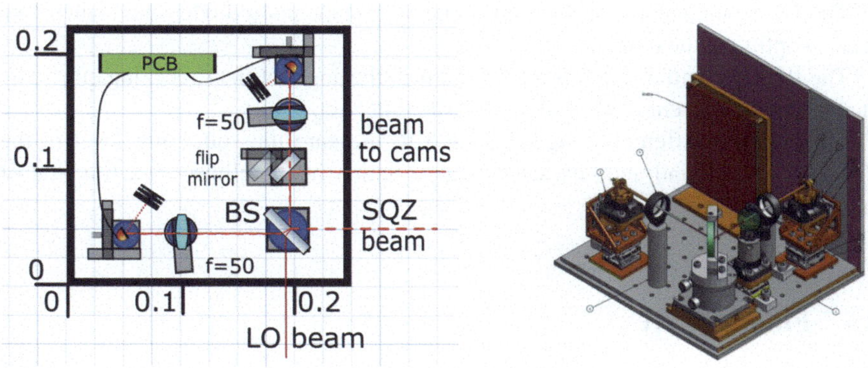

Fig. 4.8 Schematic of the external homodyne detector on EQB1. On the left the 2D optical design and on the right the CAD of the inside of the homodyne box [77]

4.3.3.1 Requirements

In order to perform precise squeezing level measurements, the homodyne needs to be well tuned in terms of contrast, Common Mode Rejection Ratio (CMRR) and Clearance.

The *contrast* is the quantity, already defined in Eq. 2.9, that determines how well overlapped the two beams are at the homodyne photodiodes, i.e. LO and the squeezing beams.

The *Common Mode Rejection Ratio (CMRR)* is the measure of the ability of the balanced homodyne to reject noises common to both photodiodes [78]. A high CMRR is necessary to subtract the classical noise present on both fields entering the beam splitter to a level below the shot noise level. An imbalance in the intensity of the beams arriving at the homodyne photodiodes leads to a worsening of the CMRR. The beam splitter has therefore to be precisely 50/50. A CMRR > 60 dB is required for good detection of squeezed states.

The *clearance* is the ratio between the homodyne dark noise (i.e. electronic noise) and the shot noise one wants to measure [78]. The clearance must be higher than the amount of squeezing one wishes to measure. The homodyne detector on EQB1 has therefore a clearance of 23 dB for detecting a maximum level of squeezing of 12 dB.

4.3.3.2 Homodyne Channels

With the homodyne, several signals can be extracted from the photodiodes: a continuous DC signal, a demodulated RF signal and a signal in the AUDIO band (i.e. from few Hz to 10 kHz). The subtraction of the signals detected by the two photodiodes gives the differential channels.

The DC signal is the total the power arriving on the photodiodes and is therefore used to optimize the contrast.

The RF differential signal is used to control the angle of the squeezing ellipse by means of the coherent control loop, described in Sect. 5.4.2.

The AUDIO differential signal is used to measure the squeezing level in the Virgo detection band and characterize the rotation of the ellipse as a function of the frequency.

4.4 Filter Cavity

The filter cavity allows the rotation of squeezed states as a function of frequency to achieve a broadband reduction of quantum noise, as shown in Sect. 3.7.2. The ellipse rotation must occur around 25 Hz, which corresponds to the frequency where the shot noise and the radiation pressure noise are equal. The parameters of the cavity, its design, the specifications of the mirrors and their suspensions, have been chosen for this purpose. The in-air filter cavity end bench (FCEB) is located in transmission of the end mirror and it hosts photodiodes and cameras used for the alignment of the filter cavity.

4.4.1 Cavity General Specifications

The filter cavity is a 285 m-long suspended Fabry-Perot cavity (see Sect. 2.1.2) with a curved-curved configuration, i.e. with both mirrors having the same radius of curvature of 558 meters. In addition, the input mirror has a wedge of 220 μm to prevent parasitic beams from entering the Faraday Isolator on SQB1 (more details in Part III). In Fig. 4.9, a picture of the filter cavity tube is shown. The optimal finesse (at 1064 nm)ok of the cavity to rotate the ellipse at the standard quantum limit (SQL)[53] is computed to be 11000 for AdV+. The filter cavity and mirrors main parameters are summarized in Table 4.3.

4.4.2 Mirrors and Suspensions of the Filter Cavity

The quality of the filter cavity mirror surfaces is the main contribution to the round-trip losses. The main specifications of the input and end FC mirrors are summarized in Table 4.4, giving both the requirements for AdV+ and the measurements performed at LMA [79, 80].

In order to obtain a residual cavity length rms motion of the sub-picometer order, some requirements on the cavity suspension system are set. Actually, a large dynamic range for positioning all the six degrees of freedom of the mirrors, a high

Fig. 4.9 Picture of the filter cavity tube, parallel to the North Virgo arm

Table 4.3 Main nominal parameters of the filter cavity [74]

Cavity parameter	Value	Mirror parameter	Value
Length L_{fc}	285 m	Material	Suprasil
Finesse at 1064 nm	11,000	Diameter	15 cm
Finesse at 532 nm	100	Thickness	9 cm
$\Delta\omega_{fc}$	$2\pi \times 25$ Hz	Wedge (FCIM)	220 μm
FSR_{fc}	526 kHz	Radius of curvature	558 m

accuracy to control locally the mirror with optical levers, an inertial dumping and seismic isolation at micro-seismic peak for scattered noise mitigation, are required. The compact suspension system is shown in Fig. 4.10. On the bottom, there is the inverted pendulum. On top of it, there is a cage hosting a double pendulum structure. This system is then inserted in a micro-tower in vacuum [74].

4.5 Suspended Squeezing Benches

The two suspended benches send the FIS produced on EQB1 to the filter cavity, called SQB1 and SQB2. Then they inject the squeezing reflected from the cavity (FDS) into the interferometer, as shown in Fig. 4.6. The goal of these benches is to mode match the beam in both directions, i.e. to the FC and to the ITF, with the telescopes they host. The infrared beam propagates from EQB1 towards the suspended bench SQB1, where it passes through a FI. Afterwards, it propagates

Table 4.4 Input and end mirrors characteristics [79, 80]. S1 and S2 stand for Surface 1 and Surface 2 of the mirrors

Filter cavity input mirror		
Parameter	AdV+ requirements	LMA measurement
Flatness RMS, S1	< 2 nm RMS	0.58 nm RMS
Radius of curvature	(558 ± 10) m	(557.9 ± 0.2) m
Average scattering at 1064 nm, S1	–	5.5 ppm
Absorption HR at 1064 nm, S1	–	0.7 ppm
Reflectivity AR at 1064 nm, S2	< 200 ppm	(4.5 ± 3) ppm
Reflectivity AR at 532 nm, S2	< 500 ppm	1150 ppm
Transmission at 1064 nm	(610 ± 80) ppm	(562 ± 1) ppm
Transmission at 532 nm	(3 ± 1.5)%	2.6%
Filter cavity end mirror		
Parameter	AdV+ requirements	LMA measurement
Flatness RMS, S1	< 2 nm RMS	0.67 nm RMS
Radius of curvature	(558 ± 10) m	(559 ± 0.9) m
Average scattering at 1064 nm, S1	–	8 ppm
Absorption HR at 1064 nm, S1	–	1.4 ppm
Reflectivity AR at 1064 nm, S2	< 200 ppm	(7.6 ± 4) ppm
Reflectivity AR at 532 nm, S2	< 500 ppm	515 ppm
Transmission at 1064 nm	< 5 ppm	(3.16 ± 0.01) ppm
Transmission at 532 nm	(3 ± 1.5)%	2.7%

towards SQB2 and then arrives at the input mirror of the filter cavity. The squeezing beam is almost totally reflected as it sees the detuned cavity. The sub-carrier beam, on the other hand, enters the cavity. The beams are then reflected from the cavity and go to the interferometer, passing again through SQB1 and SQB2. This time the beam passes through two FIs on SQB1 before being injected to the ITF. There are also cameras and actuators on SQB1 for the alignment of the squeezing system with the interferometer, sensing the ITF spurious beam. A pick-off of the reflected beam is extracted on SQB2 and reaches EQB2, where sensors for the FC length and alignment control loops with the sub-carrier are present. The optics on these benches ensure the beam to have the right properties to match the filter cavity and the interferometer. Most mirrors are motorized with picomotors, so that it is possible to align the beams remotely especially when the towers are in vacuum. In addition, some optics, mirrors or lenses, are installed on translation stages, driven by lead (Pb) Zirconate Titanate (PZT) piezoelectric actuators, to translate the optic along the beam direction. In Figs. 4.11 and 4.12, picomotors are circled in blue and PZTs in purple, for SQB1 and SQB2 benches respectively. Both benches, SQB1 and SQB2, are suspended and controlled in vacuum.

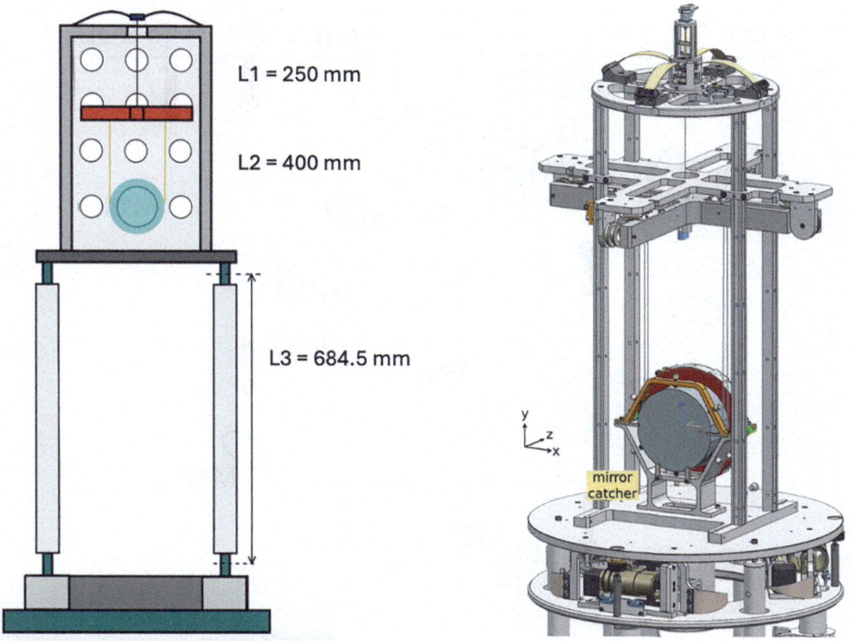

Fig. 4.10 Suspension scheme for both the input and the end mirrors [74]. On the left, the block scheme of the mirror suspension system is presented. At the bottom there is the inverted pendulum on top of which there is the actuation cage and the marionette system holding the mirror. A baffle in front of the mirror is installed to absorb scattered light. On the right, the details on the double pendulum sitting on the inverted pendulum bench is shown. The marionette actuation coils holder is attached on the cage frame. The stiffeners support on the backside supports the mirror ring heater and the coils for mirror actuation

4.5.1 Suspended Squeezing Bench 1 (SQB1)

SQB1 has dimensions 1.3×1.3 m. The optical scheme is shown in Fig. 4.11. The infrared beam in which the squeezed beam, the coherent control and the sub-carrier are superimposed arrives from EQB1 via the viewport at the top of the scheme and arrives on the first mirror SQB1_M11. It then enters the Faraday Isolator and exits towards the filter cavity at an angle of 39°. The green auxiliary beam, on the other hand, goes directly towards the cavity after the first mirror SQB1_GM11. The green beam and the infrared beam are superimposed after the mirror M21 before going towards SQB2 and the filter cavity. A green pick-off reaches two PSDs, which measure the beam position to control the bench in position as will be described in the next chapter. The infrared beam returning from the FC can be sent either towards the ITF or towards the HD by tuning the half-wave plate (HWP) between the two FIs. In the first case, the beam is transmitted by both Faraday Isolators and injected into the interferometer via the telescope consisting of SQB1_M33 and SQB1_M34.

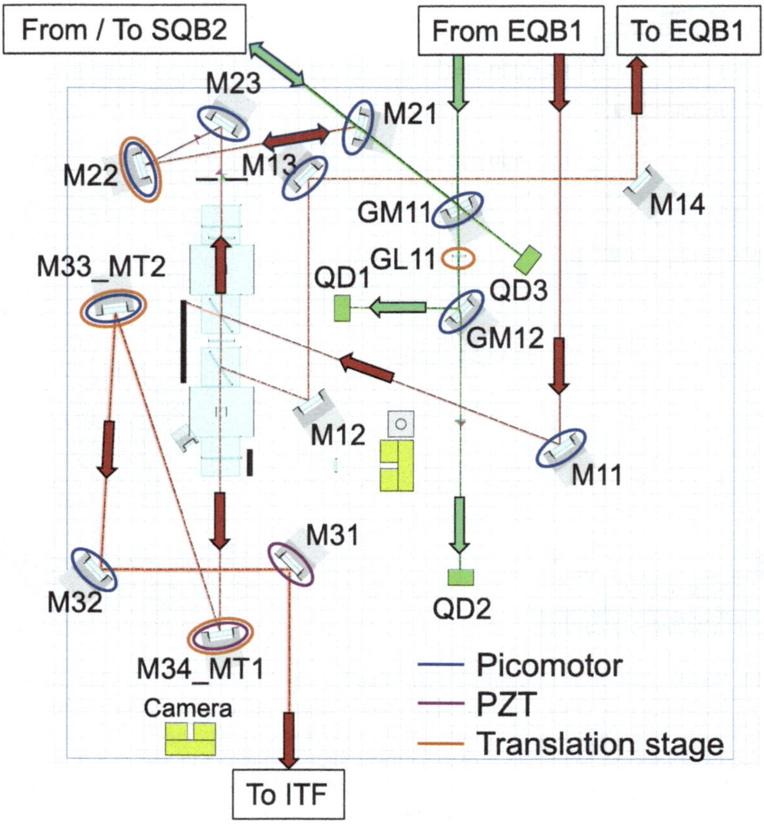

Fig. 4.11 SQB1 optical scheme with remotely controlled picomotors circled in blue, PZTs in purple and translation stages in orange

This telescope is needed to match the beam coming from the filter cavity to the ITF. In the second case, it is reflected by the second Faraday Isolator towards M21 and is directed back towards EQB1 and the HD. The green beam reflected by the filter cavity is detected by a third PSD. In addition, there are two cameras on the bench, looking at the spurious beam coming from the interferometer. This spurious beam will be used to pre-align the squeezing system with respect to the interferometer beam. Moreover, a retro-reflecting mirror is placed (not on the scheme) between M22 and M23, to send FIS to the interferometer or to the homodyne, short-cutting the filter cavity. It is mounted on vertical translation stage.

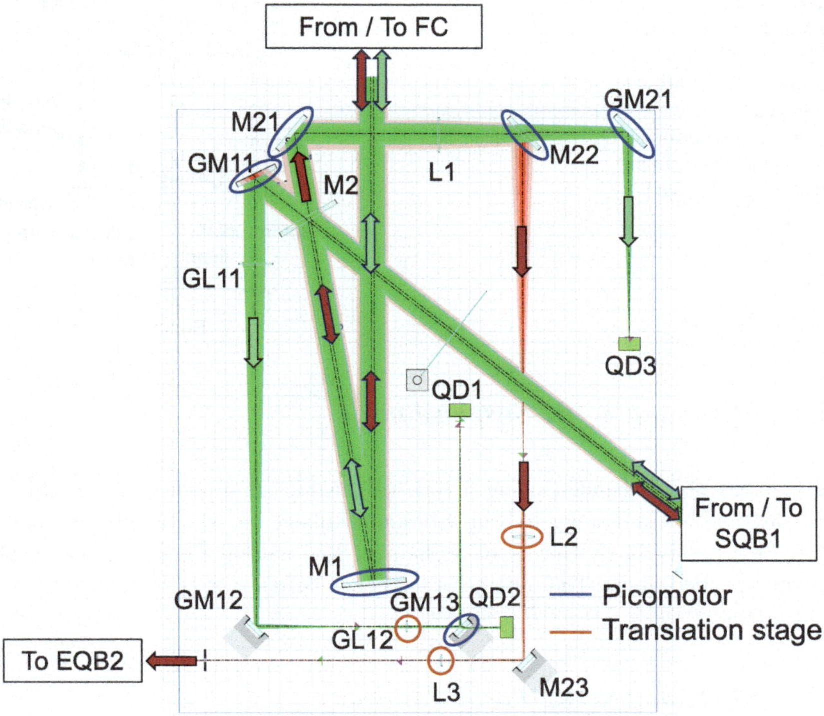

Fig. 4.12 SQB2 optical scheme with remotely controlled picomotors circled in blue and translation stages in orange

4.5.2 Suspended Squeezing Bench 2 (SQB2)

SQB2 has dimensions 1.415 × 1.225 m. The optical scheme is shown in Fig. 4.12. The green and infrared beams overlap on the last mirror of SQB1 (M21) and arrive on SQB2_M2. A pick-off of the green is transmitted and is sensed by two PSDs, one in near field and the other one in far field. The main beam goes directly into the cavity via M1. The beams reflected by the FC (both green and infrared), are partially transmitted by M2. M2 has a transmissivity of 0.1% for the infrared, which adds as optical losses for the squeezing. The green pick-off reaches the third PSD while the infrared pick-off is sent towards EQB2 and is used for control purposes.

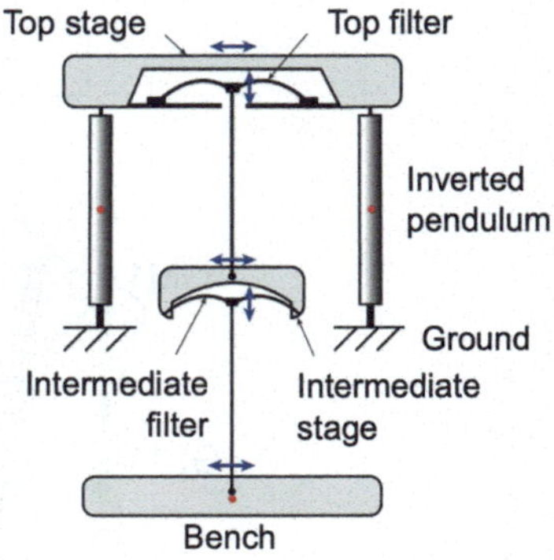

Fig. 4.13 The conceptual scheme of the MultiSAS, composed of two vertical (Geometric Anti-Spring springs) and three horizontal isolation stages (Inverted Pendulum and two simple pendulums), is shown [74]

4.5.3 *SQB1 and SQB2 Suspensions*

The benches are suspended using a type of seismic attenuator called multiSAS. It consists of a multi-stage hybrid vibration isolator in which bulk attenuation is provided passively by means of a chain of low natural frequency mechanical oscillators, called filters. The scheme of the bench suspension is shown in Fig. 4.13.

4.5.4 *SQB1 and SQB2 Local Controls*

The position control of these benches is done with pairs of coils, called Linear Variable Differential Transformers (LVDTs) [81]. For each bench, sixteen coils are required, eight of which (primary coils) are attached directly to the bench, while the other eight (secondary coils) are attached to the bench frame which sits on the bottom of the vacuum chamber. Their secondary coils are therefore connected to the ground [82]. Four pairs of coils are used to read the vertical position of the bench and the other four pairs to read the horizontal position of the bench. They are placed at the four corners of the bench. On the left side of Fig. 4.14, a photo of a pair of LVDT coils is shown. The LVDTs allow to measure the position of the bench and also to actuate on it. The reference system of the bench coordinates is shown on the right side of Fig. 4.14. The angular degrees of freedom (TX, TY,TZ) of the bench are controlled by sending a current to the secondary coils. The longitudinal degrees of freedom (X, Y, Z) of the bench are controlled using the multiSAS. The LVDTs have a displacement sensitivity of $10^{-10}\,\mathrm{m}/\sqrt{\mathrm{Hz}}$ and dynamic range of $\pm$ 1 cm [83]. To generate a good linear error signal (see Sect. 2.4.2), a modulation line at $\sim$10 kHz is sent to the primary coils. The error signal is then given by demodulation of the signal read by the secondary coil.

Fig. 4.14 On the left, a picture of a pair of LVDT coils is shown [82]. The primary coil is the one in the center and the secondary one is the cylinder containing it. On the right, the reference system of the benches coordinates is shown

4.6 Conclusion

This chapter reports the results obtained in Virgo from the injection of frequency-independent squeezed states during O3. It describes the improvements brought to reduce the losses and the introduction of the filter cavity in order to generate frequency dependent squeezed states . The different parts of the system and all the auxiliary beams used to control, characterize and align it (CC, SC, LO, BAB, GR) have been described. The next chapter will describe the commissioning of this system in Virgo.

Chapter 5
Quantum Noise Reduction System Commissioning

In this chapter, I described the work done regarding the commissioning of the quantum noise reduction system for Advanced Virgo Plus. The main work concerns the squeezing suspended benches, for which I participated in the pre-installation and pre-alignment of the optics on SQB1 and SQB2 benches at LAPP. I then participated in the installation of the two suspended benches at the Virgo site, where we suspended and controlled them inside the vacuum towers. In this context, I was in charge of testing and installing the PSD sensors used to control the suspended benches by referring to the green beam going to and coming back from the filter cavity. Part of my work was to figure out how to use these sensors into the overall control scheme of the system. Moreover, I worked on the commissioning of the entire FDS system, which started once the SQB1 and SQB2 benches were controlled, with the alignment of the green beam from EQB1 to the cavity and then reflected back onto EQB1. After that, I also participated in the alignment of the infrared beam. I contributed to the lock of the filter cavity with the green, despite it was not the main focus of my work. Once the system was aligned and the cavity locked, I worked on the optimization activities that led to the first FDS measurements with the homodyne detector. In this context, I worked on the control loops needed for these measurements, in particular the coherent control loop that controls the angle of the squeezing detection quadrature. Finally, on this system I developed a feedback loop to control the scattered light that prevented the rotation of the ellipse around 25 Hz from being seen on the homodyne, as will be described in Part III.

5.1 Pre-installation of Suspended Benches at LAPP

We installed the optics on SQB1 and on SQB2 following the nominal positions of the elements shown in the optical layout described in Sects. 4.5.1 and 4.5.2. The

E. Polini, *Broadband Quantum Noise Reduction in Advanced Virgo Plus*,
Springer Theses, https://doi.org/10.1007/978-3-031-95143-5_5

optics on the two benches allow the beam coming from the squeezer to be matched to both the filter cavity and the interferometer. To meet the required parameters of the beams, optical simulations were carried out using the *OptoCad* software [84]. For the alignment of SQB1 and SQB2, we used a collimated diode laser capable of generating both green and infrared light. For SQB1, we also performed a pre-mode matching of the beam: we reproduced the interferometer beam in order to pre-tune the telescope consisting of M33 and M34.

5.1.1 SQB1 Installation and Pre-mode Matching at LAPP

Here we describe the pre-mode matching of the M33-M34 telescope. To pre-mode match the beam and complete the installation of the optics on the bench, we followed several steps, which we describe below. The laser diode is connected with a fiber to a collimator, which directs the beam towards the optical set-up.

In the following, the beam measurements were all carried out in the same way. We placed a CMOS camera at different distances from the collimator and measured the beam size in the horizontal and vertical directions ($w_x(z)$ and $w_y(z)$) as a function of position along the beam direction z, using a software called *GxClient*. This software displays the camera images and fits the beam profile with a Gaussian function. The beam radius along the propagation direction z is defined as

$$w(z) = w_0\sqrt{1 + \left(\frac{z}{z_R}\right)^2} \tag{5.1}$$

where

$$z_R = \frac{\pi w_0^2 n}{\lambda} \tag{5.2}$$

is the Rayleigh range, w_0 is the beam waist, n is the medium refractive index and λ is the beam wavelength. By fitting the measured $w_x(z)$ and $w_y(z)$ with this function we can extract w_0 and its position z_x and z_y, as illustrated in Fig. 5.2.

1. Telescope Preparation

In order to perform the pre-mode matching of the M33-M34 telescope, we need to simulate the beam coming from the interferometer. In order to do that, we designed and installed a telescope after the laser diode. The nominal parameters of the beam we want to reproduce, i.e. the one coming from the interferometer to SQB1, are:

$$w_{0x} = w_{0y} = 1.35\text{mm} \quad \text{at} \quad z = -3.8\text{m} \tag{5.3}$$

taking as reference position the last mirror on SQB1, M31.

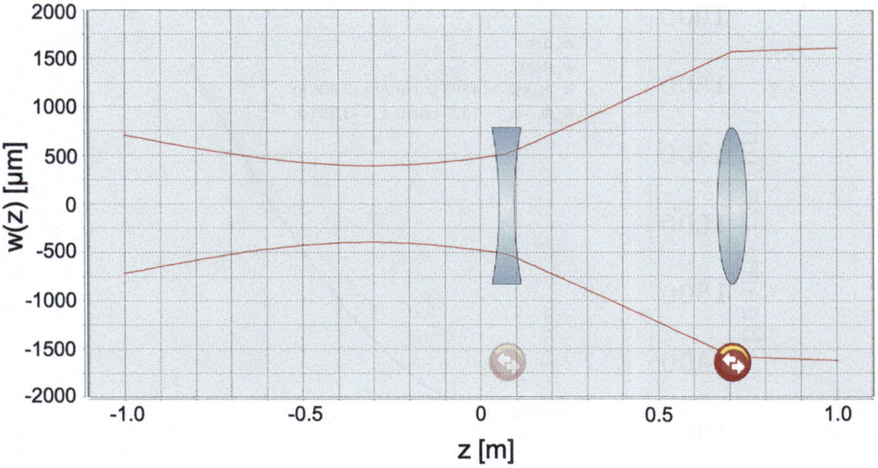

Fig. 5.1 Chosen design for the telescope; simulation on JaMmt

Using the method described above, we measured the parameters of the beam exiting the collimator and we obtained:

- $w_{0x} = 0.393$ mm at $z_x = -312.7$ mm;
- $w_{0y} = 0.392$ mm at $z_y = -317.8$ mm.

With JaMmt (Just an other Mode matching tool), a software developed in java by AEI, I simulated possible telescopes compatible with the available optics in the laboratory and the space constraints on the optical bench. I chose the following telescope design, shown in Fig. 5.1, composed by two lenses:

- Lens 1 with focal length $f_1 = -500$ mm at $d_1 = 70$ mm from the collimator;
- Lens 2 with focal length $f_2 = 1000$ mm at $d_2 = 710$ mm from the collimator

which should allow to convert the collimated beam to the nominal beam in Eq. 5.3.

2. Installation and Tuning of the Telescope

We installed this telescope and remeasured the size of the beam coming out of the telescope, setting as position zero the position of the mirror M31 on SQB1, and we obtained:

- $w_{0x} = 1.457$ mm at $z_x = -3.430$ m;
- $w_{0y} = 1.463$ mm at $z_y = -3.669$ m.

We then tuned the positions of the lenses to obtain parameters closer to those expected and positioned the telescope optics at:

- Lens 1 at $d_1 = 50$ mm from the collimator;
- Lens 2 at $d_2 = 653$ mm from the collimator.

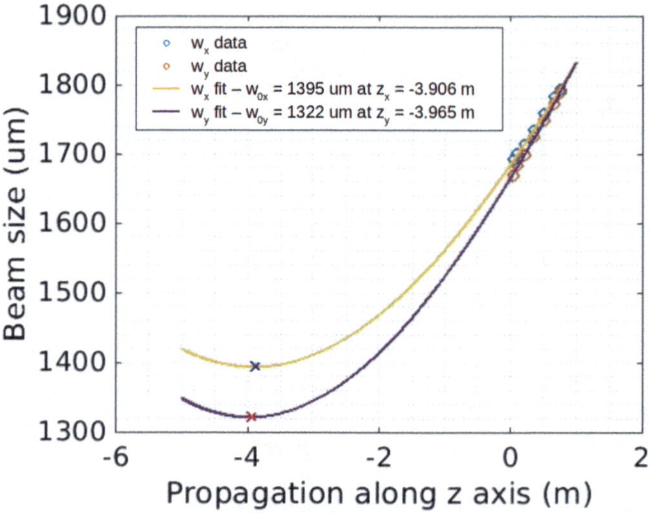

Fig. 5.2 Gaussian fit of the data collected after the telescope, the beam radius in the two orthogonal direction along the axis of propagation of the beam, and the resulting waist position and size of the beam

The beam exiting the telescope finally has the following properties:

- $w_{0x} = 1.395$ mm at $z_x = -3.906$ m;
- $w_{0y} = 1.322$ mm at $z_y = -3.965$ m.

and the fit is shown in Fig. 5.2. The results are in a good enough agreement with the expected nominal values for an initial tuning.

The matching telescope installed on the bench is shown in Fig. 5.3.

3. Installation and Alignment of ITF-SQZ Telescope on SQB1
Then we started to install and align the optics on the SQB1 bench in their nominal positions as shown in Fig. 4.11. We started by installing the mirrors M31, M32, M33 and M34. The last two mirrors are the two mirrors composing the telescope used to match the FDS beam with the interferometer. We have therefore pre-tuned this telescope. We adjusted the positions of its mirrors in order to obtain the output beam (in the filter cavity direction) matching the theoretical parameters. The beam after the M33-M34 telescope should satisfy

- $w_{0x} = 0.640$ mm at $z_x = 506$ mm;
- $w_{0y} = 0.720$ mm at $z_y = 485$ mm.

This means that the beam should have the waist right at the exit of the second Faraday Isolator. We adjusted the M33 and M34 positions in order to match the required parameters and we obtained:

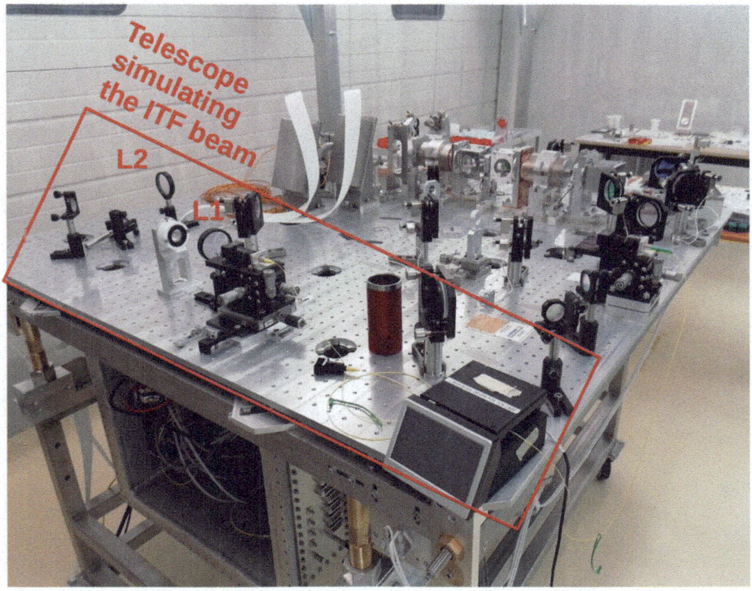

Fig. 5.3 Matching telescope installed on SQB1 bench in the LAPP clean room

- $w_{0x} = 0.594$ mm at $z_x = 681$ mm;
- $w_{0y} = 0.600$ mm at $z_y = 639$ mm

which we consider as good enough for a pre-mode matching.

4. Installation and Alignment of Faraday Isolators and Path EQB1—FC
In order to go on with the optics installation, we moved the laser diode to simulate the beam coming from EQB1. We installed the first mirror M11 which directs the beam to the first Faraday Isolator. The two FIs have been pre-aligned previously by an other team from Virgo and are mounted on a breadboard so that the relative alignment of both FI is preserved. In order to align the two FIs we sent the beam through them from the FC side. After that we checked that the beam was passing through the four optics that were previously aligned (M31, M32, M33, M34) after the double Faraday, in the ITF direction. Then we aligned the mirror M11 to have the beam reflected from the FI going to the FC direction. We adjusted a waveplate on the beam path in order to maximize the transmission of the Faraday. Then we proceeded with the installation of the following 3 mirrors on the FC side: M23, M22 and M21. The mirrors have been quite well aligned except M21 which is aiming at SQB2 and is hard to align without a reference. Moreover, we aligned the mirrors of the periscope around the Faraday Isolator in the ITF direction and we installed the 2 cameras on the bench.

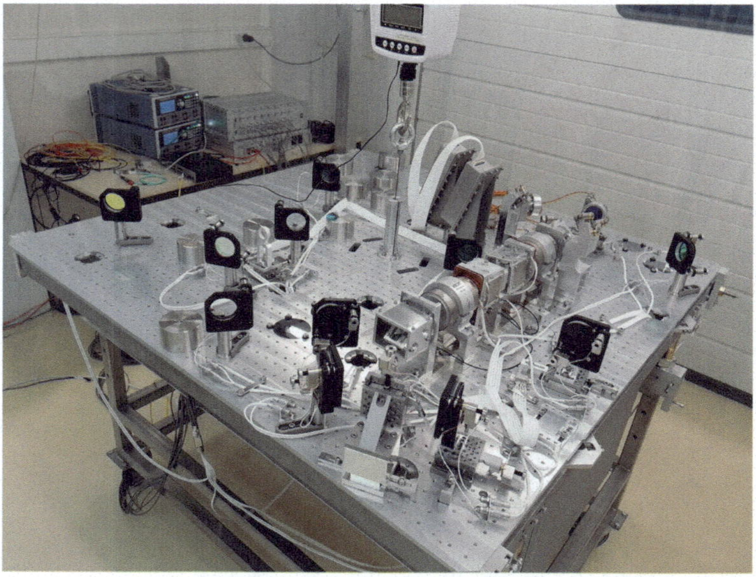

Fig. 5.4 Picture of SQB1 bench after the optics installation inside the LAPP clean room

5. Installation and Alignment of the Path to EQB1 Homodyne
We turned the half wave plate between the two Faraday Isolators to get a beam going towards EQB1. The beam was reflected by the retro-reflecting mirror, placed between M22 and M23 on a motorized vertical translation stage. We could then install the 3 remaining optics on that path: M12, M13 and M14.

6. Installation and Alignment of the Green Path
The last optics we installed are the optics on the green path. We installed and roughly aligned GM11, GM12 and the lens GL11. The PSDs were not available at the time of the installation at LAPP so they have been tested and installed directly on site later on.

7. Picomotors and PZTs Identification and Configuration
Finally, we connected and tested the picomotors and the PZTs. All of them work properly.

The bench SQB1, after the installation of all the optics described above, inside the clean room at LAPP, is shown in Fig. 5.4.

5.1.2 SQB2 Installation at LAPP

On SQB2 we didn't perform the pre-mode matching of the beam because it was not possible to generate a beam of the right dimension on the bench, as it arrives from

SQB1 on SQB2 with a radius of ~10 mm. Since there are only mirrors and lenses on this bench as shown in Fig. 4.12, the alignment was easier. Since the mode matching is not necessary, we used a green laser diode to align the optics. We positioned the laser to simulate the direction of the beam coming from SQB1. We then installed the optics on the path to the cavity, inserting M2 and M1 in that order. We then aligned the forward green pick off, installing GM11, GL11, GM12 GL13 and GM13. The PSDs were installed later on the site. We then aligned the beam in reflection from the cavity. To do this, we installed a mirror in front of M1, at the edge of the bench, to reflect the beam as if it was reflected by the input mirror of the filter cavity. We aligned the beam reflected from this auxiliary mirror so that it was superimposed to the incoming beam. Next, we installed M21, M22, and GM12. Then we completed the installation of the infrared pick-off to EQB2 with L2, M23 and L3.

5.1.3 Calibration of the LVDTs

Once the optics have been installed on the bench, we installed the LVDT pairs both on the bench and the frame. Then we calibrated their error signals in order to reconstruct the degrees of freedom of the bench in radians (angular) and meters (longitudinal). The error signal expressed in Volts, as a function of meters, can be expressed as

$$k\left(\frac{\mu\text{m}}{\text{V}}\right) = \frac{\Delta x_{LVDT}(\text{V})}{\Delta x_{LVDT}(\mu\text{m})} \tag{5.4}$$

where $k\left(\frac{\mu m}{V}\right)$ is the calibration factor. To calculate the k factor, we moved the secondary coils of each LVDT by a known amount using a screwdriver to make small, precise steps and read the relative displacement in volts from the coils. The calibration factor is computed as an average of a set of measurements. We repeated this procedure for both benches and the results are shown in Table 5.1.

Then we convert the horizontal and vertical LVDT signals into the angular and longitudinal degrees of freedom of the bench, shown in Fig. 4.14. This is done using

Table 5.1 Calibration factors for the coil pairs on SQB1 and SQB2. H stands for horizontal and V stands for vertical

LVDT Pair	$k_{SQB1}\left[\frac{\mu m}{V}\right]$	$k_{SQB2}\left[\frac{\mu m}{V}\right]$
Front Right H	81.00	79.15
Front Right V	−65.24	75.63
Back Right H	71.00	−63.70
Back Right V	87.59	72.96
Front Left H	61.20	−83.00
Front Left V	−63.12	62.59
Back Left H	66.00	65.90
Back Left V	−76.06	59.95

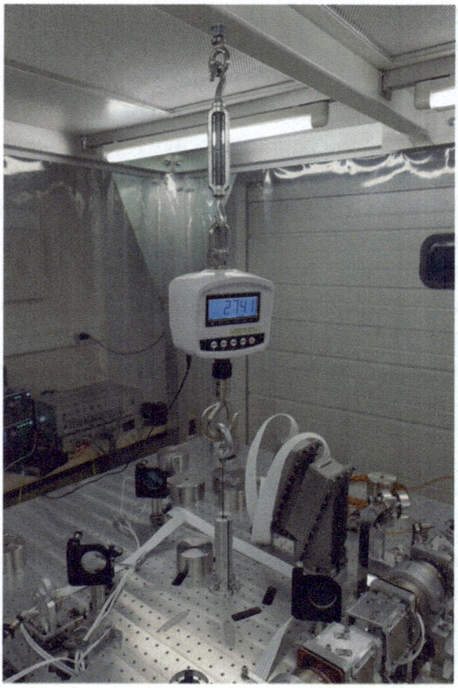

Fig. 5.5 Lifting and weighing of SQB1 at LAPP

a conversion matrix, calculated from geometric considerations of the bench and of the LVDTs. We want to control the bench at a given position, which represents the reference position of the bench. A setpoint at which we want to control the bench is therefore added to the error signal, generated by the coils as described in Sect. 4.5.4. After filtering the error signal, the correction is sent to the secondary coils to control the bench at this nominal position. The correction for longitudinal degrees of freedom is sent to the LVDTs located on the bench suspension (multiSAS), see Sect. 4.5.3.

Finally, the benches were lifted by a hook to be weighed in the clean room of LAPP, as shown in Fig. 5.5. The suspensions at the site are designed to suspend a weight of (326 ± 1) kg. To reach this weight, it has been necessary to add $\sim$50 kg for SQB1 and $\sim$100 kg on SQB2 using counter weights on the bench (cylinders of different sizes and weights) and metal plates inside the electronic box under the bench.

5.2 Suspended Benches Installation on Site

After completing the pre-installation work at LAPP, we installed the benches on site inside the vacuum chambers. In Fig. 5.6, the photos of SQB1 on the left and of the SQB2 on the right are shown. The benches are inserted inside the vacuum chambers

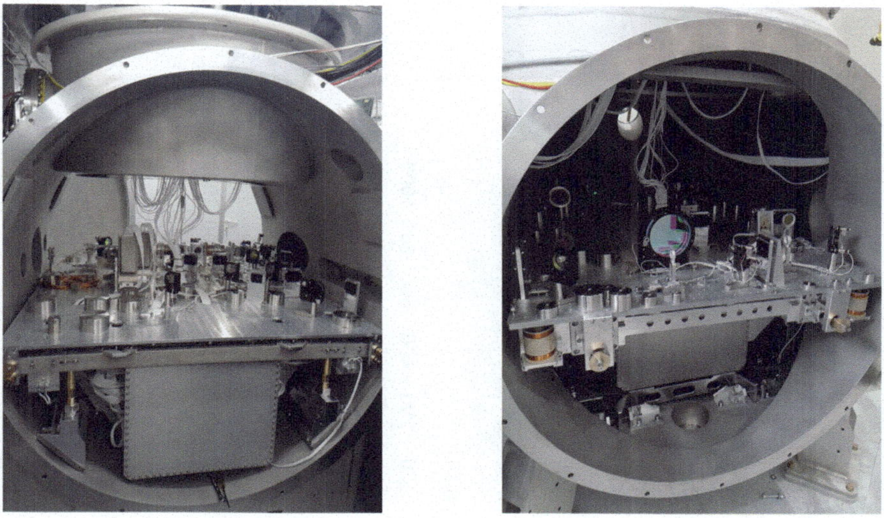

Fig. 5.6 Photo of SQB1 insalled on site on the left and of SBQ2 on the right, both inserted in their vacuum chambers

in the clean rooms of Virgo. The bench is mechanically fixed to a frame with a blocking system. The bench can be released by acting on the blocking system and it needs to be balanced before to be suspended. The bench is suspended by a wire made of maraging steel located in the center of the bench for SQB1 and slightly off-centered for SQB2. To control it in its nominal position, LVDTs are used as described in Sect. 4.5.4. In Fig. 5.6 the vertical coils of the LVDT are visible on the photo on the right: the secondary coil is attached to the frame and contains the main coil attached to the bench.

5.2.1 *Angular Control of Suspended Benches with LVDT*

We installed the wire connecting the bench to the suspension. To suspend the bench, it is needed to release its blocking system and to balance the counterweights on the bench to avoid contact with the frame. Then we engaged the bench's local angle feedback loop to control it in its nominal position. The correction sent to the magnets should be within $\pm$ 10 V. We checked that the angular degrees of freedom of SQB1 and SQB2 can be successfully controlled both in air and in vacuum with the LVDTs. As an example, in Fig. 5.7 the signals related to SQB1 angular degrees of freedom controlled in vacuum are shown: the nominal angular positions, the error signals and the corrections. The actuators can correct up to a hundred μrads in their linear range for TX and TZ and few hundreds μrads for TY. To control the balancing of the bench in TX and TZ, we can either move the counter-weights on it by hand or move

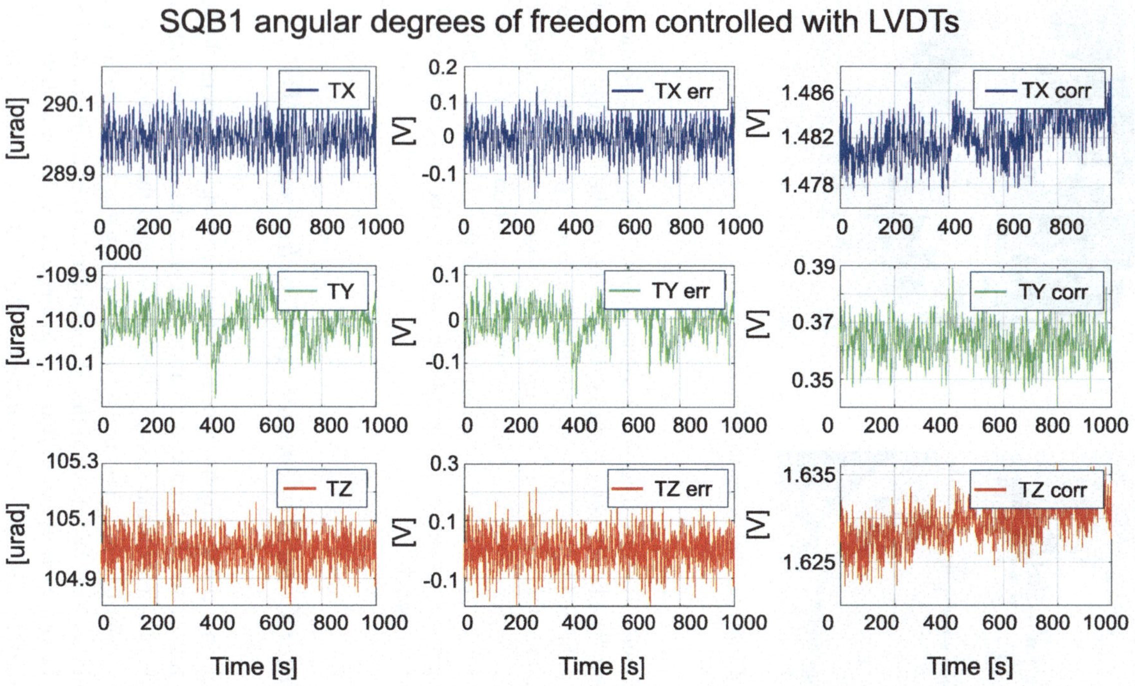

Fig. 5.7 SQB1 angular degrees of freedom (dofs) controlled with the LVDTs of the bench. Each raw corresponds to a dof: TX (blue), TY (green) and TZ (red). On the left, the controlled position; in the middle, the error signal; on the right, the correction signal

a counter-weight located on a translation stage (inside the electronic container) that we can control from remote. Finally, we also controlled the X, Y, Z positions by acting on the suspension (see Sect. 4.5.3).

5.2.1.1 Relative Residual Motion of the Suspended Benches

LVDTs are, however, referred to the ground. The benches are controlled in such a way as to minimize the residual motion with respect to the ground. Figure 5.8, shows the angular and longitudinal residual noise of SQB1 and SQB2 and the signals of the seismometer (velocimeter) in Virgo's central building (CEB), which houses the suspended benches when they register a high seismic activity. The residual angular noise of the two benches is around 1–2 μrad and the residual longitudinal noise is < 10 μm.

Figure 5.9, shows the same plots in low seismic conditions. SQB1 and SQB2 angular and longitudinal residual noises are a factor 10–20 higher in high seismic activity.

To eliminate the coupling of residual bench noise with the ground motion, we have conceived a new control system for the suspended benches, which uses the green auxiliary beam arriving on the benches as a reference. The error signal is then extracted from the position of the beam on the Position Sensor Devices (PSDs) from which it is possible to reconstruct the degrees of freedom of the bench, as will be described below.

5.2.2 Angular Control of Suspended Benches with PSDs

As part of the overall control strategy for the squeezing system, it was decided to install Position Sensor Devices (PSDs) on the suspended benches. One of the reasons was because the sensing done with these sensors refers to the green beam that impinges on them. The LVDTs, on the other hand, refer to the ground, as described in Sect. 5.2.1. The usefulness of these sensors is therefore twofold: correcting small drifts of the position of the bench controlled by the LVDTs and incorporating them into the overall control strategy of the system. PSDs are continuous area photosensors used to determine the position of the laser beam arriving on their photosensitive area. They were provided by the Institute of Cosmos Sciences (ICCUB). The model used is the S2044 of Hamamatsu. The PSD has four current outputs, expressed in X/Y coordinates (X1, X2, Y1, Y2) as shown in Fig. 5.10. A transimpedance stage is used to convert the current output signal from the PSD to voltage. The spectral response of the sensors is shown on the left side of Fig. 5.10.

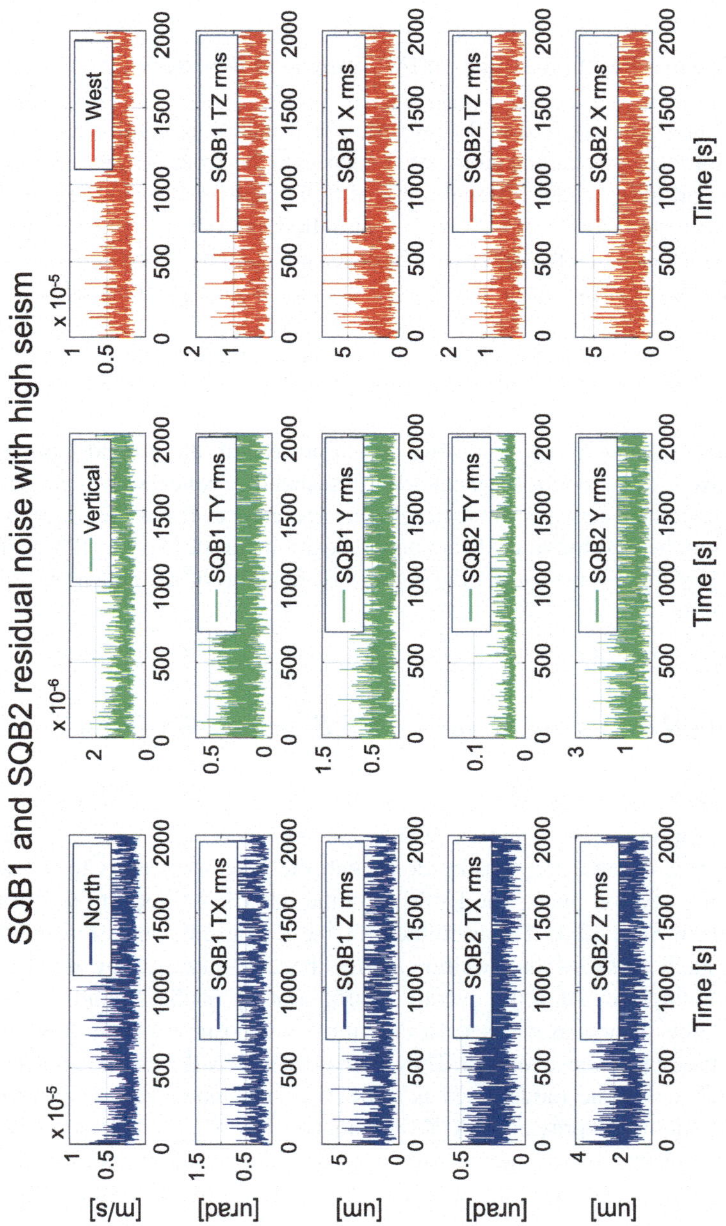

Fig. 5.8 SQB1 and SQB2 angular and longitudinal residual noise with high seismic activity

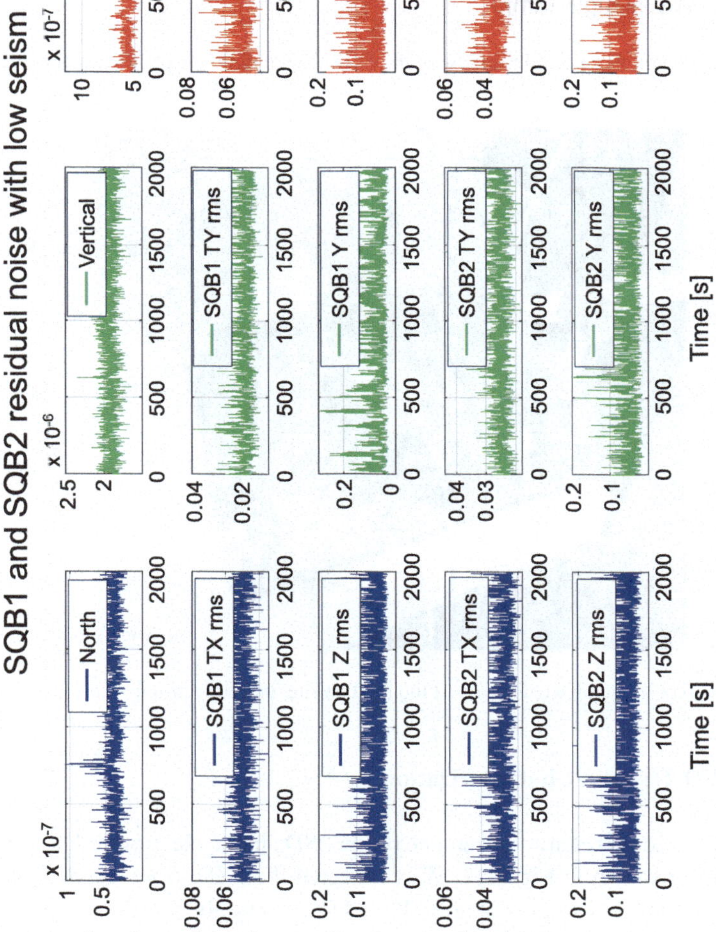

Fig. 5.9 SQB1 and SQB2 angular and longitudinal residual noise with low seismic activity

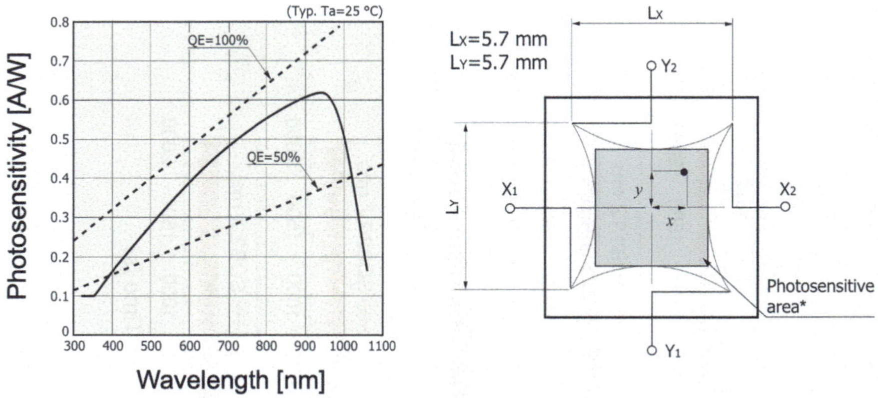

Fig. 5.10 On the left, the spectral response of the PSD. On the right, the photosensitive area chart

Fig. 5.11 Photo of the setup we used to test the PSD on site, in the electronics lab

5.2.2.1 PSD Test and Characterization

We initially tested and characterized one of the PSDs in the electronics lab in Virgo, using the setup shown in Fig. 5.11. We connected the PSD to a Data AcQuisition (DAQ) box in the lab: one pin connected to a DAC to supply it with $\pm$ 10 V and the four pins connected to an ADC.

The formulas to reconstruct the positions of the beam on the PSD from the detected currents are:

$$x = \frac{L_X}{2} \cdot \frac{(I_{X2} + I_{Y1}) - (I_{X1} + I_{Y2})}{I_{X1} + I_{X2} + I_{Y1} + I_{Y2}} \tag{5.5}$$

$$y = \frac{L_Y}{2} \cdot \frac{(I_{X2} + I_{Y2}) - (I_{X1} + I_{Y1})}{I_{X1} + I_{X2} + I_{Y1} + I_{Y2}} \tag{5.6}$$

where $L_X = L_Y = 5.7$ mm (see Fig. 5.10).

Finally, we verified that the PSD was sensitive to both horizontal and vertical beam shifts. We mounted the sensor on a small bread board, sending a fiber laser on it through a focusing lens and a steering mirror, as shown in Fig. 5.11. Then by moving the horizontal and vertical knobs of the mirror, we moved the beam first along one direction and then the other. The signals read by the PSD as a result of a horizontal (x) or vertical (y) shift of the beam are shown in Fig. 5.12. It can be seen that the horizontal to vertical coupling is $\sim$1% and that the vertical to horizontal coupling is $\sim$6%. The residual coupling between the two degrees of freedom may be due to the steering mirror in front of the sensor that makes the beam arriving not perfectly perpendicular to the PSD surface. In conclusion, the PSD works correctly.

5.2.2.2 PSD Installation on the Benches

After characterizing and testing a PSD, we installed the three PSDs on SQB1 and the three PSDs on SQB2, in the positions shown in Figs. 4.11 and 4.12. This was done

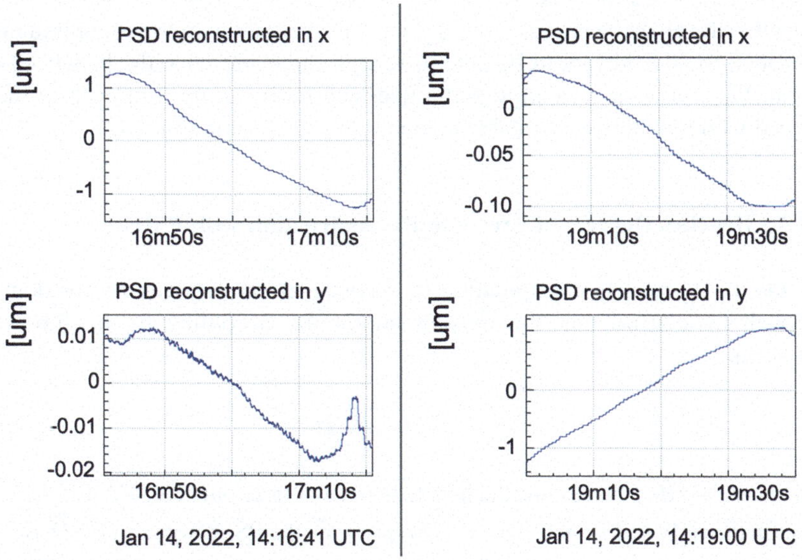

Fig. 5.12 Response of the PSD reconstructed horizontal (x) and vertical (y) signals to a shift of the beam along the horizontal direction in the left column and vertical in the right column, by acting on the steering mirror

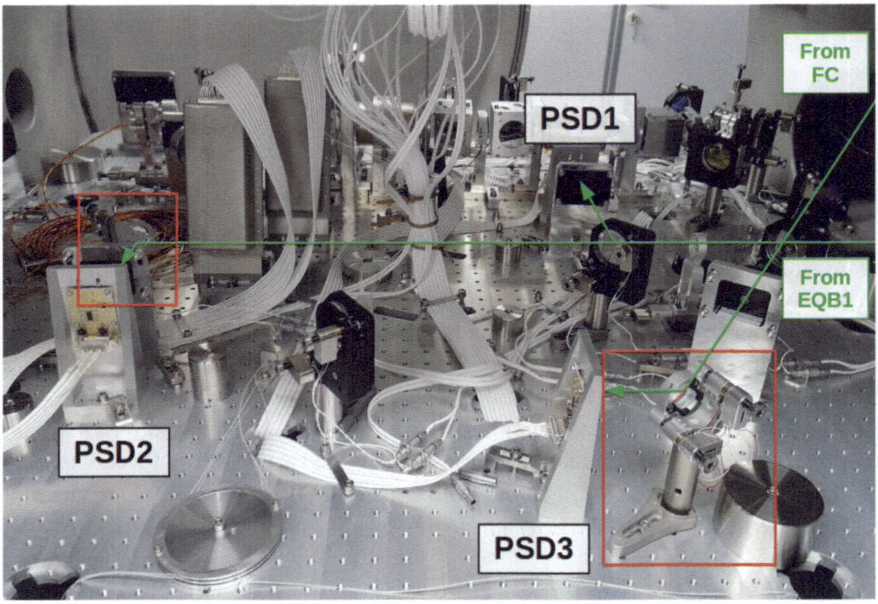

Fig. 5.13 Photo of SQB1. The two additional mirrors are indicated by the red boxes

after aligning the FDS, as will be explained in the next Section. We connected the PSDs to the DAC and ADC inside the electronic box under the suspended benches. We verified that the PSDs operate both in air and in vacuum on the benches. Due to the fact that the green beam was moving a lot at the beginning of the commissioning. we decided to add mirrors to be able to realign the beam once the bench was in vacuum. We therefore decided to add a one-inch mirror in front of each of them, equipped with remotely controllable picomotors, as shown in Fig. 5.13.

5.2.2.3 Benches Degrees of Freedom Reconstruction with PSDs

With the PSD signals, it is possible to reconstruct the degrees of freedom of the bench by constructing the sensing matrix M_s according to the following relationship:

$$\overrightarrow{dof} = M_s \cdot \overrightarrow{PSD} \tag{5.7}$$

where the vector $\overrightarrow{dof}$ represents the bench degrees of freedom (TX, TY, TZ, X, Y, Z) shown on the right of Fig. 4.14 and $\overrightarrow{PSD}$ represents the PSDs signals ($PSD_i H/V$).

The sensing matrix is calculated as the inverse of the transpose of the coefficient matrix M_c

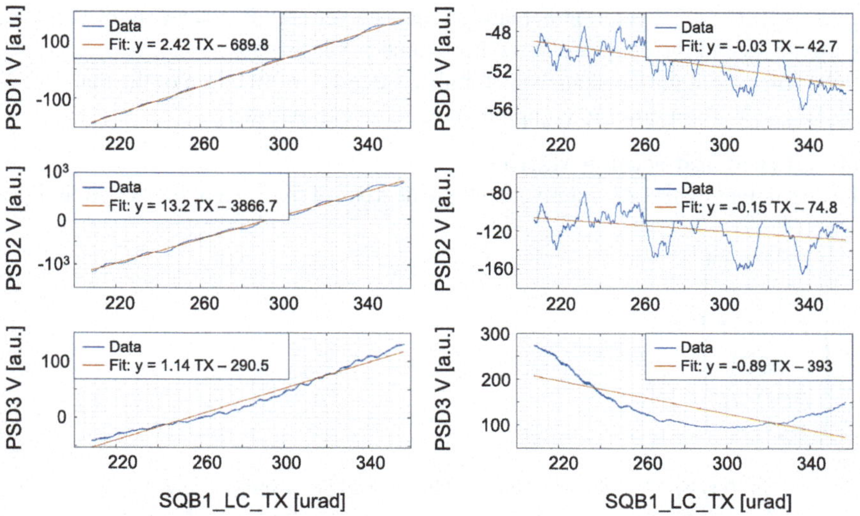

Fig. 5.14 Fit of the H and V degrees of freedom of the three PSDs on SQB1 with respect to the scan of SQB1 TX degree of freedom

$$M_s = \left(M_c^T\right)^{-1}. \tag{5.8}$$

We describe in the following the method that was followed to reconstruct the degrees of freedom of the benches, giving the example in detail of SQB1.

1. Scans of the Degrees of Freedom of the Benches
The coefficient matrix is calculated from the longitudinal and angular scans of the benches, performed using the LVDTs. We moved one degree of freedom at a time, e.g. SQB1 TX, around its nominal position. During these scans, the cavity is kept locked with the green beam since PSD3 detects the beam reflected by the filter cavity. The amplitude of the scans has been chosen in order to not misalign the beam arriving at the filter cavity too much, to avoid its unlock. Also, the scan amplitude was within the linear range of the LVDTs, in order not to lose control of the bench during the scans. The filter cavity alignment loops are open during these scans. This was done because the actuators on the green would have moved the beam arriving from EQB1.

2. Coefficients of M_c from the Fit of PSD Signals
We studied the response of the three PSDs on the bench as a function of each scan of the bench degree of freedom. For each PSDs we consider both degrees of freedom: horizontal (H) and vertical (V). If we consider as an example the scan of SQB1 TX, we fit the signal of the PSDs (H and V) as a function of the signal TX measured by the LVDTs. The fits are shown in Fig. 5.14. The slope of each fit represents a coefficient of the first row of the coefficient matrix M_c.

As expected a TX rotation is mainly seen as a vertical displacement of the beam on the PSDs and the coupling to the horizontal is very small.

We carried out the same analysis for all the degrees of freedom of the bench (TX, TY, TZ, X, Y, Z) in order to complete the coefficient matrix.

3. Coefficient and Sensing Matrices

The coefficient matrix M_c relative to all the degrees of freedom of the bench SQB1 is

$$\begin{pmatrix} PSD1H \\ PSD1V \\ PSD2H \\ PSD2V \\ PSD3H \\ PSD3V \end{pmatrix} = \begin{pmatrix} -0.030 & 2.758 & -0.295 & 1.990 & -0.027 & 0.098 \\ 2.425 & 0.015 & 0.210 & 0.050 & -1.910 & 0.050 \\ -0.150 & 14.439 & -2.407 & 16.893 & 0.013 & 0.720 \\ 13.156 & -0.362 & 1.753 & 0.105 & -16.095 & 0.139 \\ -0.891 & 2.173 & 0.379 & 0.074 & -0.190 & 0.732 \\ 1.142 & 0.048 & -0.646 & -0.038 & 0.325 & 0.104 \end{pmatrix} \cdot \begin{pmatrix} TX \\ TY \\ TZ \\ X \\ Y \\ Z \end{pmatrix}. \tag{5.9}$$

The sensing matrix is the inverse of the transpose of the coefficient matrix and from Eq. 5.9 one gets

$$\begin{pmatrix} TX \\ TY \\ TZ \\ X \\ Y \\ Z \end{pmatrix} = \begin{pmatrix} 0.069 & 1.130 & -0.010 & -0.134 & -0.048 & -0.016 \\ 0.978 & 0.072 & -0.115 & -0.011 & -0.016 & -0.031 \\ -0.194 & 2.610 & 0.013 & -0.341 & 0.116 & -1.518 \\ -0.750 & 0.315 & 0.146 & -0.040 & -0.025 & -0.225 \\ -0.015 & 1.209 & -0.001 & -0.209 & -0.015 & -0.172 \\ -2.648 & 0.092 & 0.308 & -0.005 & 1.294 & 0.838 \end{pmatrix} \cdot \begin{pmatrix} PSD1H \\ PSD1V \\ PSD2H \\ PSD2V \\ PSD3H \\ PSD3V \end{pmatrix}. \tag{5.10}$$

4. Offline Reconstruction of the Bench dofs from M_s

Before actually using the values for the sensing of the benches, which would have required machine time, we simulated the reconstruction of the degrees of freedom

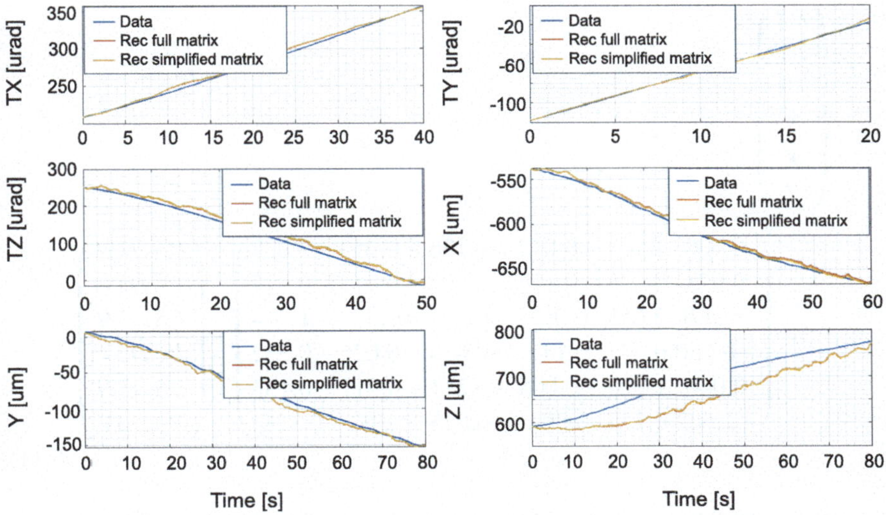

Fig. 5.15 SQB1 six degrees of freedom reconstruction with the complete (red curve) and the simplified (yellow curve) matrix, i.e. with small coupling factors set to zero, overlapped with the LVDT data in blue. The angular dofs are expressed in μrad and the longitudinal ones in μm

of the benches with *Matlab*. We used the sensing matrix to reconstruct the bench scans. We reconstructed the degrees of freedom of SQB1 using the matrix Eq. 5.10 and multiplying it by the PSD signals during the scans used for the fits, as described in Eq. 5.7. For example, SQB1 TX degree of freedom is properly reconstructed if reproduce the displacement measured by the LVDT during the scans.

The reconstructions of the degrees of freedom of SQB1, using the complete matrix in Eq. 5.9 are shown by the red curves in Fig. 5.15. The bench movement detected by the LVDTs at the same time is shown by the blue curves.

Normally, a movement of a vertical degree of freedom of the bench, i.e. TX, TZ or Y, gives a vertical displacement of the beam on the PSD. Similarly, for a scan of a horizontal degree of freedom of the bench, i.e. TY, X or Z, we see a horizontal displacement of the beam on the PSD. This is evident from the fits in Fig. 5.14, where by moving the bench in TX, we see the beam moving mostly vertically on the PSDs.

An interesting test is to see if it is possible to improve the reconstruction removing the couplings between the H and V degrees of freedom of each PSD. We decided to set to zero the coupling factors < 5% for each PSD and to consider only those > 5%.

The complete sensing matrix, recalled M_A in Eq. 5.10 would become:

$$\begin{pmatrix} TX \\ TY \\ TZ \\ X \\ Y \\ Z \end{pmatrix} = \begin{pmatrix} 0.100 & 1.143 & -0.012 & -0.136 & -0.050 & -0.027 \\ 0.978 & 0.025 & -0.115 & -0.003 & -0.015 & -0.027 \\ -0.176 & 2.655 & 0.017 & -0.348 & 0.112 & -1.544 \\ -0.750 & 0.347 & 0.147 & -0.046 & -0.026 & -0.232 \\ 0.040 & 1.226 & -0.005 & -0.211 & -0.017 & -0.183 \\ -2.602 & 0.225 & 0.303 & -0.025 & 1.290 & 0.821 \end{pmatrix} \cdot \begin{pmatrix} PSD1H \\ PSD1V \\ PSD2H \\ PSD2V \\ PSD3H \\ PSD3V \end{pmatrix}. \tag{5.11}$$

and the reconstructions obtained from this matrix are shown by the yellow curve in Fig. 5.15. It can be seen that for some degrees of freedom, e.g. X, the reconstruction is slightly better. For the others, the improvement is not very evident. This is why we have derived reduced coefficient and sensing matrices, to be able to decouple the degrees of freedom of the bench better.

5. Reduced Coefficient and Sensing Matrices
We have decomposed the sensing matrix in Eq. 5.10 into smaller matrices, thus reconstructing less degrees of freedom of the bench with less PSD signals. We performed several Matlab reconstruction simulations to check which of the reduced matrices best reconstructed the degrees of freedom. Since we only control the angular degrees of freedom of the bench with the LVDTs, we ensured to reconstruct the angular degrees of freedom TX, TY and TZ in the best possible way. After analyzing and studying different combinations, we verified that they worked. Two examples of simplified sensing matrices, to reconstruct the dofs of SQB1 are given below:

- Sensing matrix M_B: reconstructing TX, TY, X, Y with PSD1 and PSD2

$$\begin{pmatrix} TX \\ TY \\ X \\ Y \end{pmatrix} = \begin{pmatrix} 0 & 1.159 & 0 & -0.136 \\ 0.946 & 0 & -0.111 & 0 \\ -0.809 & 0 & 0.154 & 0 \\ 0 & 0.947 & 0 & -0.175 \end{pmatrix} \cdot \begin{pmatrix} PSD1H \\ PSD1V \\ PSD2H \\ PSD2V \end{pmatrix} \tag{5.12}$$

  whose reconstructions with Matlab are shown in Fig. 5.16;
- Sensing matrix M_C: reconstructing TX, TY, TZ, X with PSD1 and PSD2

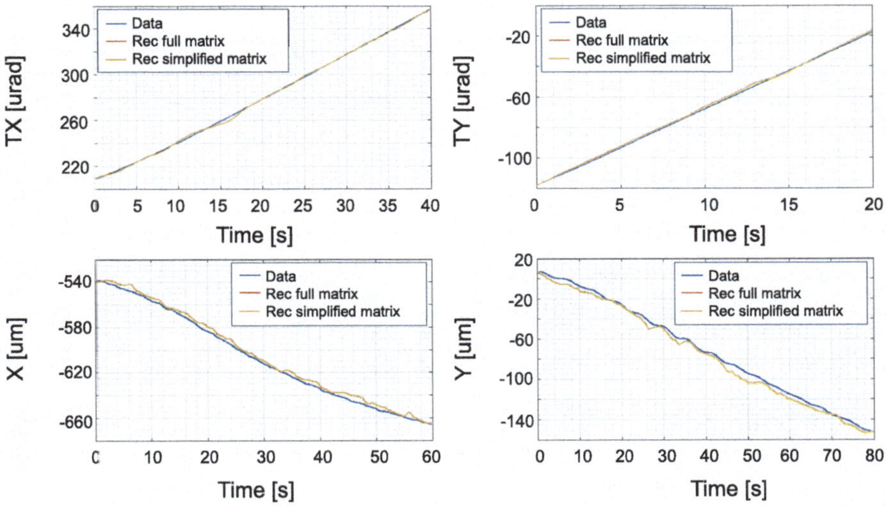

Fig. 5.16 SQB1 four degrees of freedom reconstructed (TX, TY, X, Y) with the complete (red curve) and the simplified (yellow curve) sensing matrix M_B. The angular dofs are expressed in μrad and the longitudinal ones in μm

$$\begin{pmatrix} TX \\ TY \\ TZ \\ X \end{pmatrix} = \begin{pmatrix} 0 & 1.182 & 0 & -0.142 \\ 0.946 & -0.094 & -0.111 & 0.017 \\ 0 & -8.870 & 0 & 1.634 \\ -0.809 & -1.184 & 0.154 & 0.218 \end{pmatrix} \cdot \begin{pmatrix} PSD1H \\ PSD1V \\ PSD2H \\ PSD2V \end{pmatrix} \tag{5.13}$$

whose reconstructions with Matlab are shown in Fig. 5.17.

In this case only PSD1 and PSD2 are used, so only the green beam going to the filter and not the one reflected.

6. Sensing of the Bench dofs with Different Matrices
Once the matrices were inserted into the interface used for the control of the benches, the *Virgo processing monitoring*, we checked which degrees of freedom were actually well reconstructed. To do this, we scanned the bench and saw if the sensing with LVDTs and PSDs were equivalent. We then selected the matrix for each degree of freedom that would best reconstruct it. In Fig. 5.18, reconstructions of the angular degrees of freedom of the SQB1 bench are shown using the M_A, M_B and M_C matrices (given in Eq. 5.11, 5.12 and 5.13 respectively). The first line of plots gives the values read from the LVDTs while the other plots represent the dof reconstructions. From this analysis, it can be concluded that TX and TY are well reconstructed with the M_B matrix. The degree of freedom TZ is not reconstructed well in any case, the only matrix that reconstructs its movement is M_A, but it is very noisy. This is due to the geometry of the system, as TZ is the angle around the beam propagation axis who arrives from EQB1. Therefore PSD1 and PSD2 are not very sensitive to a displacement of the bench in TZ.

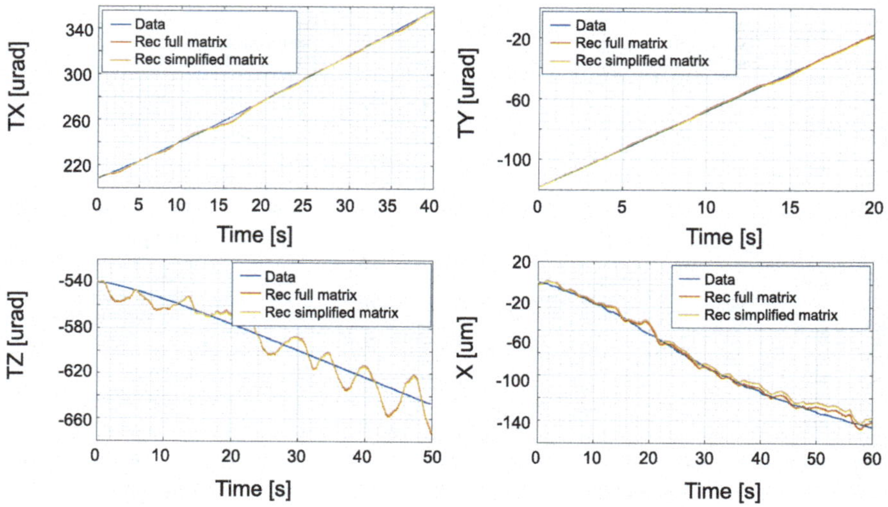

Fig. 5.17 SQB1 four degrees of freedom reconstructed (TX, TY, TZ, X) with the complete (red curve) and the simplified (yellow curve) sensing matrix M_C. The angular dofs are expressed in μrad and the longitudinal ones in μm

PSD3 is more sensitive as it sees the beam arriving from the FC which is not along TZ, but it is at 39°.

5.2.2.4 Attempt to Control SQB1 TX and TY dofs with PSD Signals

Once we verified that it is possible to reconstruct the SQB1 TX and TY degrees of freedom with the matrix M_B, we tried to use these signals as sensing in the angular control loop. The error signal is the sum of the signal given by the LVDTs (at high frequency) and the one given by the PSDs (at low frequency). The correction is sent to the coils of the LVDT as usual. We measured the transfer function of the loop by injecting noise in the TX signal reconstructed with the PSD. We found an UGF $\leq 10^{-3}$. However, we found a conflict with the cavity alignment loop on the green as it acts on the mirrors on EQB1. So if this loop is active, it misaligns the beam too quickly on the SQB1 PSDs and they cannot follow it. This causes the bench control loop corrections to diverge. We decided to switch off the cavity alignment loop and test the operation of the SQB1 loop using LVDT and PSD sensing. In Fig. 5.19 are shown the signals of the degree of freedom TX controlled in this way. In blue are shown the value of TX and the value of the correction sent to the coils to keep the bench in the desired position. Shown in green are the data read from the vertical degrees of PSD1 and PSD2, used for TX reconstruction. In red are shown the signals from the seismometers on EQB1 which measure the displacement of the bench. It can be seen that what is corrected in this way is the movement of the beam due to the displacement of EQB1 which is not suspended and subject to drifts. There

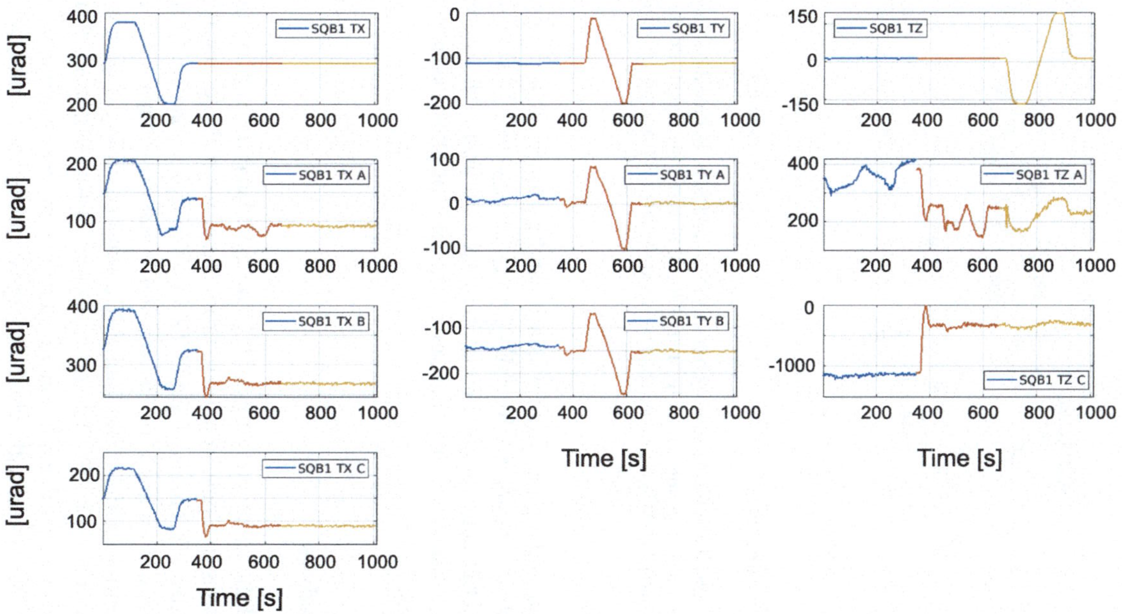

Fig. 5.18 Reconstruction of the angular degrees of freedom of the SQB1 bench with the signals from PSD1 and PSD2 using the matrices M_A, M_B and M_C. The first raw shows the signals read by the LVDTs. The other raws show the reconstructions with the M_A, M_B and M_C matrices. In blue are indicated the data during the bench scan in TX, in orange during the bench scan in TY and in yellow during the bench scan in TZ

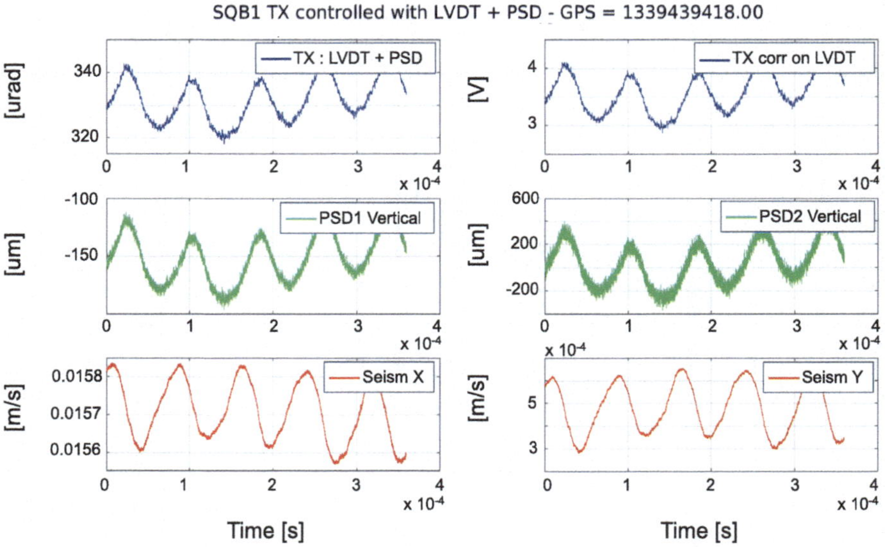

Fig. 5.19 SQB1 TX controlled with the sensing from LVDTs and PSDs. In blue the signals of the TX value on the left and its correction on the right are shown. In green the signals on the vertical signals of PSD1 (left) and of PSD2 (right) are shown. In red the signals from the seismometer on EQB1 are shown

are 10 hours of data in the plots, from which it can be deduced that the period of the oscillations is about 2 hours. So the PSDs correct, as expected, for slow drifts of the beam but the EQB1 beam is not a good reference as it drifts. The use of these signals is still under discussion and alternative strategies of global system control are being considered.

5.2.2.5 SQB2 Analysis

We also performed the same study for SQB2. We calculated the sensing matrices and studied the reconstructions on matlab. From these, we realized that it is only possible to reconstruct TY but not TX and TZ in a decoupled manner. In fact, the incident beam arrives at an angle of 39° on SQB2 and this couples the TX and TZ degrees of freedom on the PSD1 and PSD2 sensors. In Fig. 5.20 are shown the reconstructions of the TX, TY, TZ and X degrees of freedom using PSD1 and PSD2 on SQB2. It is evident that TX and TZ cannot be reconstructed using signals from these sensors. An option with which the signals are better reconstructed is to use PSD3 on SQB1, which reads the beam arriving from the cavity on a different axis. Otherwise, it might be possible to reconstruct a degree of freedom that is a combination of TX and TZ, which would correspond to rotating the reference axis.

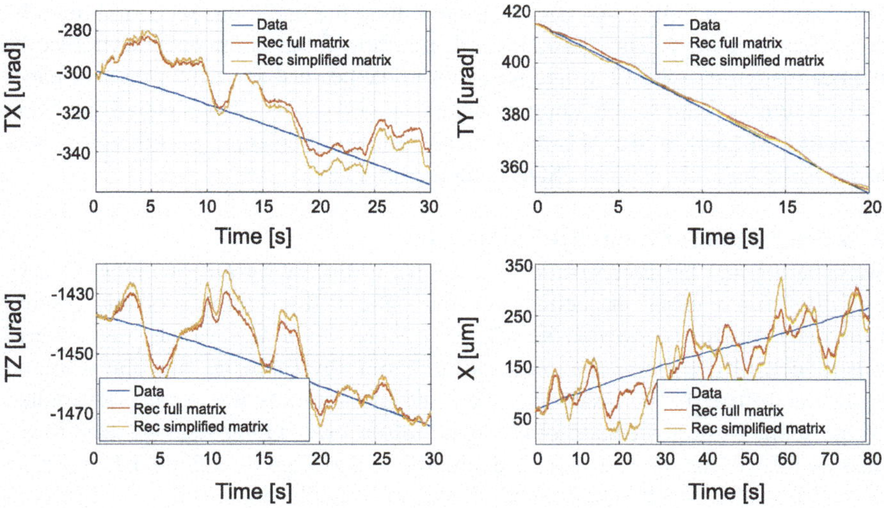

Fig. 5.20 Reconstruction of SQB2 TX, TY, TZ and X degrees of freedom using the signals from PSD1 and PSD2 on SQB2. TY is well reconstructed while TX and TZ are coupled due to the geometry of the system

5.3 FDS System Alignment

The installation of the whole FDS system (EQB1, suspended benches and filter cavity) was completed with SQB2 in April 2021. Subsequently, we controlled the vacuum benches with LVDTs as described in Sect. 5.2.1. We then started the alignment of the green and infrared beams over the entire system, from EQB1 to the end bench located in transmission of the cavity (see Fig. 4.5). In the following we describe the procedure used to complete the alignment of the system shown in Fig. 4.6 and in more details in Figs. 4.11 and 4.12.

5.3.1 Auxiliary Green Beam Alignment

We started by aligning the green beam as it follows a less complex path and it is used for the first stage of the filter cavity lock, on which the infrared beam will then be aligned.

1. Green Alignment from EQB1 to SQB2

The green beam exits the squeezer box and propagates through the optics on EQB1. It then reaches the viewport between EQB1 and SQB1, which is now open and controlled in-air. We moved the last mirror on EQB1, called M7, to center the beam on SQB1 mirrors: GM11 and GM12. The beam is therefore sent towards SQB2 from GM11. We moved this mirror with its picomotors in order to make the green beam arriving on SQB2. We saw the beam hitting the baffle inside the linking tube (see Chap. 7) by eye. This gave us an idea on how to align the beam. Afterwards, the beam arrived to SQB2, where we checked the beam shape with a sheet of paper and we removed the visible clipping acting on GM11. We also centered the beam on SQB2_M2 and SQB2_M1, the 6 inches mirrors. The beam was therefore sent towards the link tube between SQB2 and the filter cavity.

2. Green Alignment from SQB2 and FCIM
We then aligned the green beam from SQB2 to the cavity mirror input (FCIM). Initially, from a visual inspection, we saw some diffused light, probably coming from one baffle inside the pipe (see Chap. 7). Moreover, it was evident that the beam was reflected by the link tube in the east direction (by analyzing how the scattered light was diffused). The observable we could use remotely was a camera installed on the south-western viewport of the input mirror tower. Otherwise, we went to see directly by eye from the viewport. We initially tried to align by acting on the position of the suspended bench SQB2, along TY i.e. horizontally. Moving it 8 mrad was not enough so we decided to move the bench back to its original position and move the 6-inch mirror SQB2_M1. By manually steering SQB2_M1 while looking at FCIM from the sout-western viewport, we eventually found the beam hitting the mirror cage, as shown on the left picture of Fig. 5.21. However, trying to center the beam on the FCIM, we found that it was clipped vertically, on the upper part of the pipe. We then started to move SQB2_M2, adjusting SQB2_M1 accordingly, to prevent the beam from hitting the input mirror cage. We realised that this is possible by having the beam very low on SQB2_M1, as shown on the right picture of Fig. 5.21. We then investigated the problem and discovered that the bench had been installed a few centimeters too high with respect to its nominal position. We measured the height of SQB2 in the Virgo Reference System (VRS) with a theodolite optical level. The bench was (16 $\pm$ 1) mm too high. We have therefore lowered the SQB2

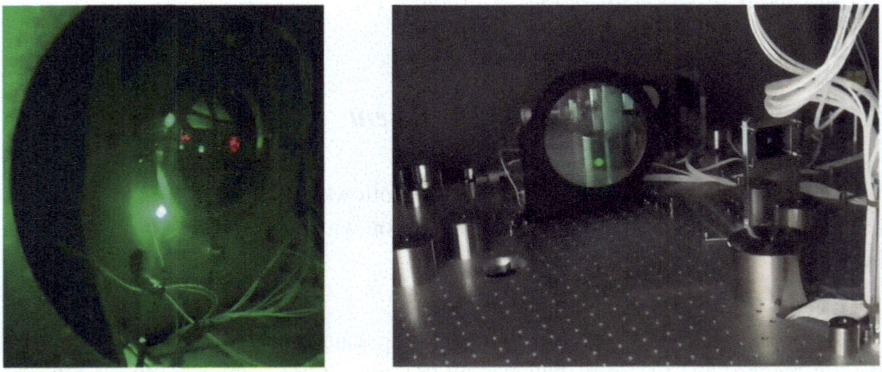

Fig. 5.21 On the left the green beam hitting the FCIM cage. On the right the green beam low on SQB2_M1 when centered on FCIM

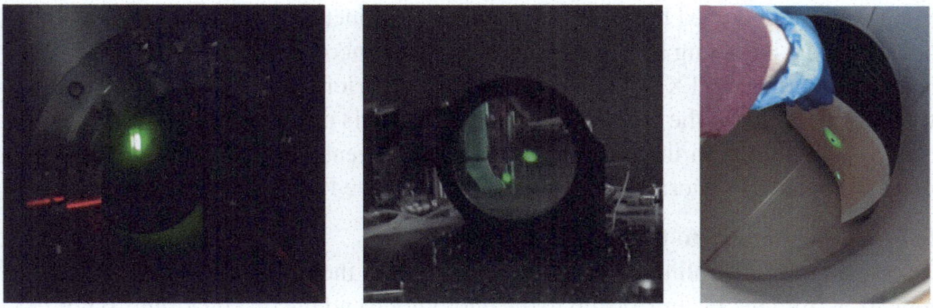

Fig. 5.22 On the left the green beam hitting the FCEM cage, seen from the south-eastern viewport. In the middle, the green beam retro-reflected from the FCIM seen on SQB2_M1. On the right, the FCEM reflection in the tube in front of the micro-tower

bench by 17.25 mm by acting on the length of the frame's four legs. We then aligned the green beam so that it was centred on both SQB2_M1 and FCIM.

3. Green Alignment from FCIM to FCEM
Once the beam arrives centered on FCIM, we need to center it on FCEM. We tuned the angle of SQB2_M1 to steer the beam while looking at the inner face of FCEM through the south-eastern viewport of its micro-tower. We immediately saw the green beam on the mirror cage, as shown on the left of Fig. 5.22. We aligned the beam on the centre of FCEM and we roughly aligned the beam on the terminal bench to reach the camera.

4. FCIM Reflection Alignment Until SQB2
To complete the alignment to the input mirror, we need to overlap the green beam reflected from the FCIM with the incoming green beam. This optical procedure is called autocollimation. The cavity mirrors are locally controlled. The suspended benches are controlled in air. We scanned the FCIM angles, both TX and TY, in order to find the back-reflection of the green beam on SQB2. The range allowed by the optical levers ($\sim$10 mrad) was not sufficient to find the retro-reflection. We left FCIM uncontrolled and free to oscillate more to increase the scanning range and we saw the retro-reflection appear on the bench. In order to go on, it was possible to open the tube in front of the FCIM input mirror in order to better understand the position of the reflected green beam. Rotating FCIM, we superimpose the incoming and the reflected beam a few tens of cm away from the input mirror, with a precision of a few mm, thanks to the help of a metal ruler with a hole to let pass the incoming beam. We could see the reflection from the FCIM on SQB2_M1, as shown in the middle of Fig. 5.22.

5. FCEM Reflection Alignment
With the FCIM misaligned, we started the alignment of the autocollimated beam on FCEM. We scanned the angles of the FCEM, to try to find the reflection on SQB2. We didn't find any reflection so we decided to open the flanged port about 3 m before the end micro-tower to find the position of the reflection. We tried to autocollimate the beam as we did for the input: letting the incoming beam pass through a hole on a piece of paper and find the reflection on the other side, as shown on the right of

Fig. 5.22. The reflected beam was a couple of centimeters far in both direction with respect to the incoming beam. We superposed the incoming and the reflected green beams by acting on TX and TY of the FCEM marionetta. After several attempts, we managed to send the beam towards SQB2. At this point we aligned this beam as best as we could with the incoming beam and we realigned FCIM as well. At this point we could see interferences on the camera at the FC transmission, on FCEB.

6. FC Reflection up to the Sensors on EQB1
Once we have autocollimated the green beam with the cavity, we have aligned the reflection from the FC mirrors up to the photodiode and the two quadrants on EQB1. The last ones are used to lock the cavity on the green.

7. Filter Cavity Locked on the Green Beam
Once the system is aligned and the longitudinal lock loop on green is ready, we lock the cavity (see Sect. 2.4). The green beam is modulated at 5.5 MHz using an Electro-Optical Modulator (EOM) on EQB1. The reflected beam from the cavity is detected by the photodiode on EQB1 that demodulate the 5.5 MHz signal, see Sect. 2.4.2. The actuation is sent to the coils of the cavity end mirror, in order to displace the mirror along the cavity axis direction. At the beginning the residual cavity length fluctuation, i.e. the lock precision, was 50 Hz rms. The unitary gain frequency of the loop is $\sim$100 Hz.

5.3.2 *Infrared Beam Alignment to the Filter Cavity*

Once the cavity is locked on the green, we have a reference on which to align the infrared beam. The OPO is set in BAB mode, i.e. instead of squeezing, there is an infrared beam resonating within it, called the Bright Alignment Beam (BAB). The sub-carrier beam (SC) is also aligned in the OPO, so aligning the BAB or SC is equivalent. We started by aligning the BAB which has a power of 1 mW and then we checked the alignment of both beams in the cavity.

1. BAB Overlapped with the Green Beam on SQB1
We started the alignment with the benches controlled in-air. We sent the BAB from EQB1 to SQB1, aligning via the mirrors on EQB1 and the mirror on SQB1, the beam with the first Faraday Isolator. The BAB should overlap with the green beam on the dichroic mirror M21. We acted on the mirrors before M21 in order to center the BAB on the mirror. Then we acted on M21 in order to overlap it with the green beam towards SQB2.

2. BAB and Green Beams Superposition into the Filter Cavity
By acting on SQB1_M22, we overlapped the green and the BAB also on SQB2_M2 and M1. Then, by acting on SQB1_M21 we overlapped the two beams at the end of SQB2, right before the linking tube towards the FCIM. Iterating the steps, we managed to have the BAB (seen with a IR viewer) well aligned with the green beam at the open flange before FCEM.

3. BAB Reflection Alignment on SQB2 and EQB1

Since the BAB is superimposed with the green beam, we were able to immediately see the reflection on SQB2, despite it was not perfectly aligned with the green beam. We started by autocollimating the BAB from FCIM, keeping the cavity aligned on the green beam. We aligned the reflection by using SQB1_M21 and M22 to get the reflection out of the Faraday Isolator and tuned the waveplate to send it towards EQB1. We centered the beam on M12, M13 and M14 on SQB1. The alignment was difficult because M12 and M14 do not have picomotors. The beam was then aligned on two cameras and one photodiode on EQB1.

4. BAB and SC Alignment in Transmission from the Cavity

On EQB1, there are remotely controllable shutters to prevent beams from propagating into the filter cavity:

- Green shutter: to stop the green auxiliary beam;
- IR shutter: to stop all the IR beams (SC, BAB/squeezing);
- SC shutter: to stop the SC beam;
- LO shutter: to stop the local oscillator beam.

We aligned both BAB and BAB + SC on the cavity by moving SQB1_M22 and M21 to maximize the flashes in transmission from the cavity. The SC beam had a power of 25 mW. In order to find the TEM00 (defined in Appendix A) of both the SC and the BAB we slowly scanned the frequency of the green beam with an Acousto-Optic Modulator (AOM) on the green EQB1 path, while the filter cavity was locked on the green. This results in a change of the cavity length. The higher order modes displayed on the camera on FCEB give us an idea of the type of misalignment or mode mismatch of the infrared beam in the cavity. Initially only modes of high order are visible, but acting on the IR mirrors on SQB1 (M21, M22 and M23) the lower order modes are found. The free spectral range scans for the green and BAB in the cavity are shown in Fig. 5.23.

5. Aligning the SQB1 Retroreflector Mirror

Next we aligned the retro-reflector mirror on SQB1, placed between M22 and M23. This mirror is mounted on a vertical translation stage and has picomotors. We checked that its reflection arrived at the same position on the cameras and had the maximum power on the photodiode by acting on its picomotors.

6. Alignment of the Sensors on EQB2

We aligned the optics and the sensors for the filter cavity on EQB2, when the cavity was stably aligned with the IR beam.

7. Lock of the Filter Cavity with the IR Beam

Using the signals from the photodiode on EQB2, we performed the hand-off of the cavity lock from the green to the SC beam. The SC is modulated at 11 MHz using an EOM in front of the laser source on EQB1. A pick-off of the SC beam reflected from the filter cavity arrives on the photodiode on EQB2. The signal is then demodulated generating an error signal. This error signal must be the same as the green error signal used for the cavity lock. The correction signal is sent to the filter cavity end mirror's coils to act on the longitudinal direction.

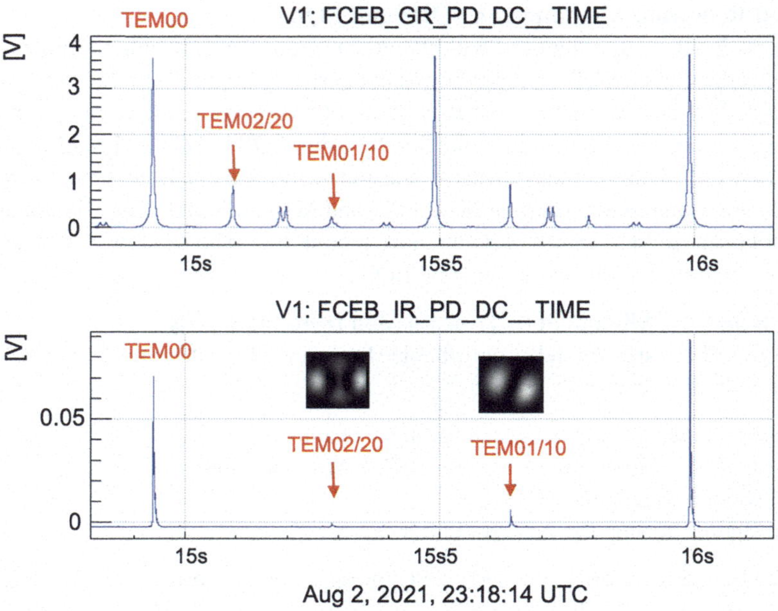

Fig. 5.23 Free Spectral Range of the green (top) and the BAB (bottom) in the filter cavity. The TEM01/10 represents the misalignment and the TEM02/20 represents the mode mismatch (more details in Appendix A)

5.3.3 Conclusion on System Alignment

Once the whole FDS system was installed and the optical components pre-aligned, we performed the global alignment as described above. The main difficulty was to align the beams on such a large distance, of the order of hundreds of meters. This work took a few months. Once the alignment of the system was completed, it was possible to finalize all the control loops that allowed us to perform the FIS and FDS measurements on our system, as will be described in the next paragraph.

5.4 FIS Measurements Preparation on EQB1 Homodyne

Once the system was aligned and the cavity locked on the green, we started squeezing measurements on the homodyne detector on EQB1. We started with the frequency independent squeezing measurements to characterize the losses and try to optimize the system before proceeding with the frequency dependent squeezing measurements. We can make FIS measurements in two different ways. The first way is with FIS propagating to the back-reflecting mirror on SQB1 and then being reflected towards the homodyne on EQB1. The second way is FIS propagating only on EQB1 through a delay line. The delay line is an alternative optical path on EQB1 that mimics the propagation of the beam on the whole system. Two one-inch mirrors, mounted on a horizontally translating, remotely controllable slide, reflect the beam onto EQB1 towards the homodyne detector. This slide is placed in front of the viewport between EQB1 and SQB1. These two detection configurations are useful to characterize squeezing losses separately for EQB1 and SQB1.

Two fundamental ingredients are needed to make FIS measurements: alignment of the squeezed field with respect to the local oscillator and the control of the quadrature angle at which we want to measure squeezing. These processes are automated by means of control loops: automatic alignment of the homodyne (HD AA) and the coherent control loop (CC loop).

5.4.1 Homodyne Automatic Alignment

In order to perform squeezing measurements on the homodyne detector on EQB1, it is necessary for the squeezing beam and the local oscillator beam to be perfectly superposed. A misalignment between them results in optical read-out losses (see Sect. 3.4). The squeezed beam is superimposed on the coherent control (CC) beam. Pre-alignment is initially done through two cameras on EQB1, one on the near-field and one on the far-field. A mirror mounted on a horizontal, remotely controllable slide (HD_MIR) in front of one of the homodyne photodiodes (see Fig. 4.3.3), reflects the beam and sends it to the cameras. It is possible to perform this alignment either with the OPO set in squeezing mode or in BAB mode. In squeezing mode we can superimpose either the CC and the LO fields, either the SC and LO fields (since the SC is brighter than the CC and matched with the OPO). In BAB mode we directly overlap the BAB with the LO on the cameras. Next we sent the two overlapped beams on both homodyne photodiodes, PD1 and PD2. These are mounted on a triaxial stage that allows them to be remotely moved horizontally, vertically and along the beam direction, in order to finely align the beam over their sensitive area. The alignment between CC and LO is done by maximizing the interference amplitude between the two beams measured with the RF difference channel (4 MHz) of the homodyne detector (see Sect. 4.3.3). The amplitude of the interference between the two fields is calculated from the two quadratures of the signal used for the CC loop (see Sect. 5.4.2) demodulated at 4 MHz (see Sect. 2.4.2). In order to maximize this quantity which indicates the superposition of the two beams, we introduce *dither lines*[1] on the mirrors on the path of the CC. These modulations are sent by a PZT on the horizontal (TX) and vertical (TY) degrees of freedom of the M4 and M6 mirrors, which are located in front of the homodyne. Their frequencies are 412 Hz, 512 Hz, 612 Hz and 712 Hz. By demodulating the

[1] *Dither lines* are sinusoidal modulations applied to the actuators around the working point, with the aim of extracting an error signal for a control loop.

amplitude of the interference between the LO and CC at the dither line frequencies we obtain four error signal. The four error signals derived from the demodulation at the angular dither frequencies are filtered generating four correction signals. They are sent to the TX and TY degrees of freedom of the M4 and M6 mirrors, controlled with PZT of dynamic range of $\sim 30\ \mu$m and bandwidth of $\sim$100 Hz. This alignment can be optimized by minimizing the four error signals which means maximizing the amplitude of the interference between LO and CC, by acting on TX and TY of the two mirrors.

5.4.2 *Coherent Control Loop*

In order to generate squeezed vacuum states with a certain orientation and detected in a given quadrature, it is necessary to introduce phase controls. We call them *coherent controls*. These are explained in this section.

5.4.2.1 Theory of the Coherent Control Loop

We have already introduced the coherent control field (see Sect. 4.2.3), which is a phase-shifted with respect to the main squeezer laser. This field is a single sideband field at a frequency of $\omega_0 + \Omega$, where ω_0 is the squeezer laser frequency and Ω = 4 MHz is the frequency shift. The expectation values of the upper and lower sideband annihilation operators of this field, following the formalism described in Sect. 3.5, are

$$\hat{a}_+ = \hat{a}_{\omega_0+\Omega} = \alpha_\Omega \qquad \hat{a}_- = \hat{a}_{\omega_0-\Omega} = 0 \tag{5.14}$$

where α_Ω represents the real value of the single sideband field. From these relations the two-photon operators $\hat{a}_1$ and $\hat{a}_2$ defined in Eq. 3.65 are obtained. The propagation of this field through the OPO, corresponds to applying the squeezing operator $S(r, \phi)$ defined in Eq. 3.32 to these states. The output states have the following amplitudes

$$\hat{b}_1(\Omega) = -i\frac{\alpha_\Omega}{\sqrt{2}}\sin(2\phi)\sinh(r) + \frac{\alpha_\Omega}{\sqrt{2}}[\cosh(r) + \cos(2\phi)\sinh(r)] \tag{5.15}$$

$$\hat{b}_2(\Omega) = \frac{\alpha_\Omega}{\sqrt{2}}\sin(2\phi)\sinh(r) - i\frac{\alpha_\Omega}{\sqrt{2}}[\cosh(r) - \cos(2\phi)\sinh(r)]\,. \tag{5.16}$$

Expressing in the time domain by means of a Fourier transform these operators, the coherent control field exiting the OPO $E^{CC}(t)$ can be expressed as [85]

$$E^{CC}(t) \propto \langle\hat{b}^{(+)}(t) + \hat{b}^{(-)}(t)\rangle \tag{5.17}$$

where

$$\hat{b}^{(\pm)}(t) = \frac{1}{2}\left(\hat{b}_1(t) \pm \hat{b}_2(t)\right) e^{\mp i\omega_0 t}. \tag{5.18}$$

This expression can be simplified by substituting $e^r = \sqrt{g}$ and obtaining [85]

$$E^{CC}(t) \propto \frac{1+g}{\sqrt{2g}}\alpha_\Omega \cos(\omega_0 t + \Omega t) - \frac{1-g}{\sqrt{2g}}\alpha_\Omega \cos(\omega_0 t - \Omega t - 2\phi). \tag{5.19}$$

Equation 5.19 shows how the coherent control field is (de-)amplified by the non-linear process occurring in the crystal. The field coming out of the OPO has two sidebands at $\omega_0 \pm \Omega$. From this expression, it is possible to extract the angle of the squeezing ellipse ϕ produced in the OPO. The error signal is given by the demodulation of the interference between the two sidebands in reflection from the OPO that contains the difference in phase between the green pump and the CC field. The loop correction is sent on a phase shifter on the green path before the cavity inside the AEI box.

Finally, the phase difference between the LO field and the squeezed field on the homodyne must be stabilized. This phase depends on the phase between the green pump field and the LO field, denoted by Φ. The LO field arriving at the beam splitter of the homodyne can be expressed as [85]

$$E^{LO}(t) \propto \alpha_{LO} e^{-i\omega_0 t} e^{-i\Phi} + c.c. \tag{5.20}$$

and is superimposed on the CC field coming out of the OPO described in Eq. 5.19. The two fields arrive superimposed on the PD1 and PD2 photodiodes of the homodyne with the following field amplitudes [85]

$$\begin{aligned} E^{PD1,PD2}(t) =& \frac{1}{\sqrt{2}}\left(E^{LO}(t) \pm E^{CC}(t)\right) \\ \propto& \frac{1}{\sqrt{2}}\Bigg[\alpha_{LO} e^{-i\omega_0 t} e^{-i\Phi} \pm \\ &\left(\frac{1+g}{\sqrt{2g}}\alpha_\Omega \cos(\omega_0 t + \Omega t)\right. \\ &\left.- \frac{1-g}{\sqrt{2g}}\alpha_\Omega \cos(\omega_0 t - \Omega t - 2\phi)\right)\Bigg] + c.c. \end{aligned} \tag{5.21}$$

The interesting signal is the difference of the two photocurrents $I^{PD1,PD2} = |E^{PD1,PD2}|^2$ which is (removing the high frequency signal at 2Ω):

$$I_{DIFF} = I^{PD1} - I^{PD2} \propto \frac{2\sqrt{2}\alpha_{LO}\alpha_\Omega(-1+g)}{\sqrt{g}} \cos(\Omega t + 2\phi + \Phi). \tag{5.22}$$

Demodulating this photocurrent at the frequency Ω, one finds the error signal for the relative phase Φ between pump green and LO fields [85]:

$$S_{err}^{CC/LO} \propto \frac{\sqrt{2}\alpha_{LO}\alpha_{\Omega}\,(-1+g)}{\sqrt{g}}\sin\left(2\phi+\Phi\right). \tag{5.23}$$

Since the squeezing angle ϕ is controlled acting on the optics located on the green path before the OPO, the LO and CC relative phase Φ can be controlled using this signal by acting on different PZT actuators along the LO path and/or the squeezing path as described below.

5.4.2.2 Experimental Coherent Control Loop

The FDS system in homodyne detection configuration is shown in Fig. 5.24, where the squeezing propagates to the filter cavity and then back towards the homodyne detector on EQB1 where it combines with the local oscillator. The same principle

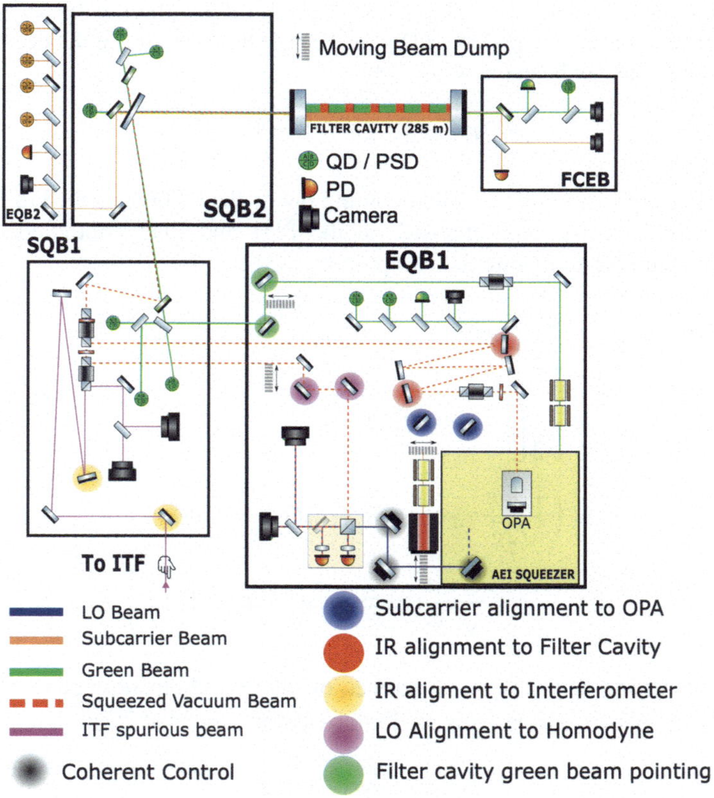

Fig. 5.24 EQB1 homodyne detection scheme. Credit: M. Vardaro

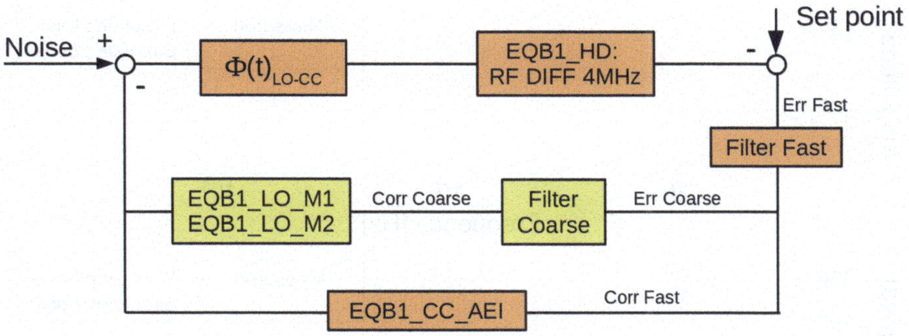

Fig. 5.25 Coherent control Nested Loop scheme

is applied to the beam that is reflected by the back-reflector mirror on SQB1 or the delay line on EQB1.

Here we describe the coherent control loop used to fix the angle of detection, i.e. Φ, of the squeezing ellipse on the homodyne. In order to have the right bandwidth and to increase the range of actuation, it was necessary to design a Nested loop, as shown in Fig. 5.25.

This loop consists of a first loop called *CC fast*, which operates with a bandwidth of $\sim$1 kHz but with no sufficient range. Therefore, an additional loop (CC coarse) is used to keep the correction of the fast CC around zero, increasing the dynamic range of the loop. The plant of the loop is the difference in phase between the LO and CC fields $\Phi(t)_{LO-CC}$ of Eq. 5.23. The sensor used is the differential homodyne signal demodulated at the radio-frequency (RF) $\Omega = 4$ MHz. Depending on the quadrature of the squeezing we want to detect we introduce a set point,[2] i.e. a fixed relative phase between the two fields Φ which coincides with the angle of the detected ellipse. The error signal obtained from the difference of the detected signal and the set-point is filtered by the CC fast filter and a part of the signal becomes the correction that is sent to an actuator on the LO path (EQB1_CC_AEI). The dynamic range of this actuator is $\sim$9 μm and its bandwidth $\sim$10 kHz. The other part of the CC fast correction signal becomes the error signal of the CC coarse loop and the purpose is to keep it around zero. This is filtered by the Filter Coarse. The coarse correction is then sent to two actuators on EQB1 on the LO path: EQB1_LO_M1 whose dynamic range is $\sim$3 μm and bandwidth $\sim$1 kHz, and EQB1_LO_M2 whose dynamic range is $\sim$5 μm and bandwidth $\sim$100 Hz.

The loop was characterized by measuring its transfer function (TF), from which the unitary gain frequency (UGF) is obtained (see Sect. 2.4). To do this, the loop was closed on the delay line, so that only the CC fast is sufficient to correct the plant. The CC coarse does not change the UGF as it has an effect at low frequencies. To

---

[2] This is not a real set point since the loop is closed in zero. We act on the demodulation phase to keep the error signal in a linear region.

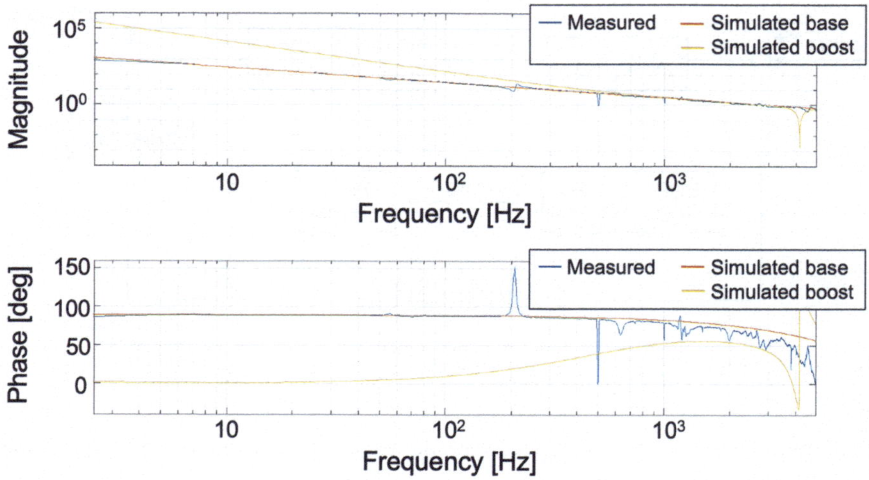

Fig. 5.26 Fast Coherent Control loop transfer function. The red curve represents the simulated TF of the loop with base filter while the yellow curve with the boost filter, i.e. with higher gain under the UGF frequency. The blue curve represents the measured TF with the base filter. The UGF of the CC loop is at 3 kHz

calculate the TF of the CC fast, we injected a noise with amplitude of 50 μV onto the PZT of the mirror in the squeezer. We measured a UGF of 3 kHz as shown in Fig. 5.26. This is an UGF higher than before (1 kHz) which is a good result. We also implemented an alternative filter that has higher gain at low frequency, which we call *boost filter*. The simulation of the boost is shown in the yellow curve of Fig. 5.26. This loop allows to fix the phase of the squeezing angle and therefore to characterize the squeezing level, both for FIS and FDS.

5.5 The Story of FIS Measurements

The first FIS measurements on the homodyne have been performed in February 2021. The measurements were repeated in the summer of 2021 before and after the OPO was replaced. The crystal used in O3 is degraded due to long use and therefore cannot produce the necessary level of squeezing for Virgo. Table 5.2 shows the measurements before and after replacing the OPO both with the homodyne inside the squeezer and with the homodyne on EQB1 using the delay line. Between the two homodynes there is therefore only the propagation on EQB1 as a difference. But from the very little amount of squeezing measured on EQB1 HD, it has been found out a 39 ± 5% of optical losses and a phase noise of 10 mrad. The contrast of the homodyne was 95% during these measurements. The reason for such high losses turned out to be due to several factors:

Table 5.2 Squeezing and anti-squeezing measurement with the homodyne inside the AEI squeezer and the EQB1 one, before and after the replacement of the OPO

Homodyne	Squeezing (dB)	Anti-Squeezing (dB)
Old OPO		
AEI squeezer HD	3.4	6
EQB1 HD	1.7	5
New OPO		
AEI squeezer HD	10.1	13.6
EQB1 HD	3.5	11.6

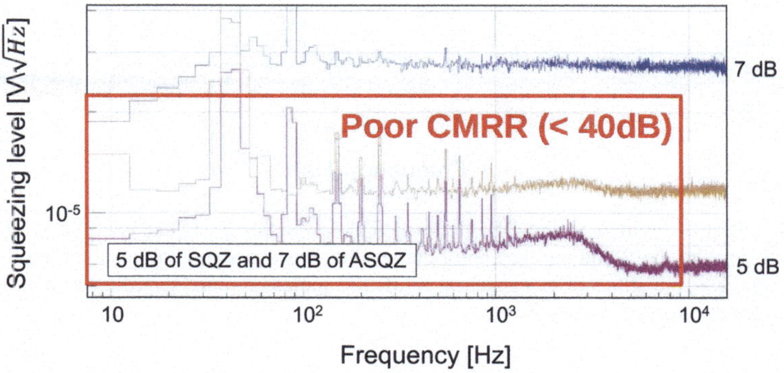

Fig. 5.27 FIS measurement from EQB1. The brown curve represents the shot noise level, the magenta curve represents the squeezing level and the blue one represents the anti-squeezing level

- initially, to align the homodyne, two mirrors on different paths were used: one on the LO and one on the squeezing path. This misaligned the beam on the HD beam splitter, changing the splitting ratio, and on the HD photodiodes, reducing the contrast (see Sect. 4.8);
- polarization losses introduced by a mirror on the delay line before the homodyne;
- the beam splitter of the homodyne was a commercial one and not the one coated at LMA so it was unbalancing the homodyne photodiodes, reducing the CMRR (see Sect. 4.3.3).

We then resolved some of these issues before proceeding with the subsequent squeezing measurements. Homodyne alignment is done either by hand or using the auto-alignment loop described in Sect. 5.4.1 with the PZTs on the M4 and M6 mirrors both on the squeezing path. The delay line mirror that introduced polarization losses was replaced.

A second measurement campaign was carried out in autumn 2021, measuring again FIS from the delay line. The contrast of the homodyne was 99%. We were able to measure 5 dB of squeezing and 7 dB of anti-squeezing above 5 kHz, but for lower frequencies, large structures can be seen that worsen the measurement and an increase in shot noise and squeezing level was observed between 1 and 5 kHz, as shown in Fig. 5.27. This effect is due to the low CMRR during the measurement, i.e. CMRR < 40 dB. To solve the issue we replaced the beam splitter of the homodyne

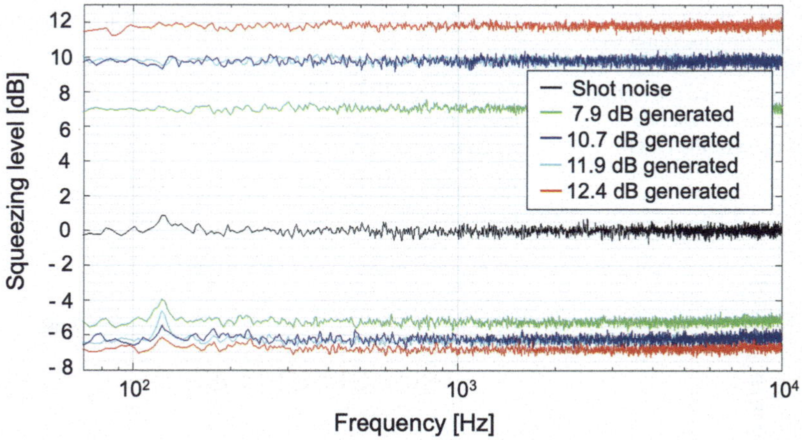

Fig. 5.28 FIS measurement from SQB1 performed in November 2021. The black curve represents the shot noise level in dB and for each color the level of anti-squeezing (above the shot noise) and squeezing (below the shot noise) measured in dB are indicated, corresponding to the generated squeezing level indicated in the legend

with an other one having a splitting ratio very close to 50/50. After this action we reached a CMRR > 60 dB.

The third squeezing measurement campaign was performed shortly afterwards, this time measuring FIS reflected by the retro-reflector mirror on SQB1. We carried out the measurements with different levels of squeezing generated by the OPO, changing the power of the green pump beam point by point. The measurements made with different levels of squeezing generated in the OPO are shown in Fig. 5.28: the black curve represents the shot noise and for each color the level of anti-squeezing (above the shot noise) and squeezing (below the shot noise) measured are indicated. The spectra of these measurements are flat down to 50 Hz, below that scattered light spoils the measurement, as it will be explained in Sect. 5.6.

By acquiring points for different green pump powers, it is possible to more accurately estimate optical losses and phase noise by fitting the squeezing level as a function of the anti-squeezing level using the expected dependence shown in Figs. 3.8 and 3.9. From the fit shown in Fig. 5.29, the estimated optical losses are 15.4 ± 0.6 % and the phase noise is 24.3 ± 10.0 mrad.

The expected optical losses are:

- 6% in the propagation on SQB1 both ways (from power measurements);
- 4–5% from the HD contrast;
- 1% from the HD photodiodes efficiency;
- 2% from the propagation on EQB1 and the OPO escape efficiency;
- 1% from dark noise (LO operated at 25% of its maximum power)

which sums up to (14.5 ± 0.5)% and agrees with the squeezing measurement.

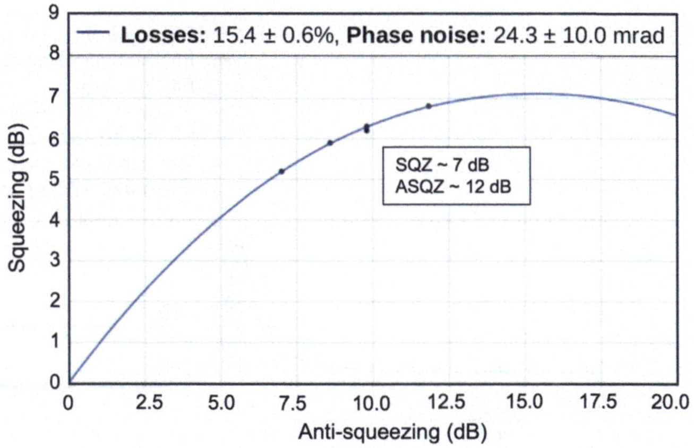

Fig. 5.29 FIS measurement from SQB1. Squeezing level as a function of the anti-squeezing level for different values of the green pump power

After this measurement we decided to send the squeezing directly to the filter cavity, which is locked with the green auxiliary beam, to measure the rotation as a function of the detuning frequency of the cavity.

5.6 First Demonstration of Squeezing Ellipse Rotation

We characterized first the filter cavity performances: lock precision and round trip losses. The lock precision is the rms value of the residual frequency noise of the cavity and is calculated from the error signal used for the lock: either the signal of the SC or the projection on the IR if the cavity is locked with green. It has been obtained an IR length noise of 8 Hz locking on IR and 11 Hz length noise on the green beam. The round-trip losses (RTL) of the cavity were measured with the reflection measurements of the SC beam on EQB2 by comparing the reflected power when the cavity is locked or unlocked [73]. It has been measured RTL of (31.0 ± 1.5) ppm, with expected value of 40 ppm.

We then moved the retro-reflecting mirror to bring the squeezed field up to the filter cavity, which is locked with green since we cannot keep the SC beam on during squeezing measurements on the EQB1 homodyne. In this way we obtained the first measurements of the rotation of the squeezing ellipse at 361 Hz, shown in Fig. 5.30. From this measurement, the estimated optical losses are 21%: 6–8% due to HD homodyne, 13–15% are propagation losses. The estimated phase noise is 30 mrad.

The two limitations of this measurement are:

- the fact that rotation cannot be measured at 25 Hz due to the low-frequency noise;
- the anti-squeezing curve does not reach the shot noise level in correspondence of the rotation hole by 1.3 dB.

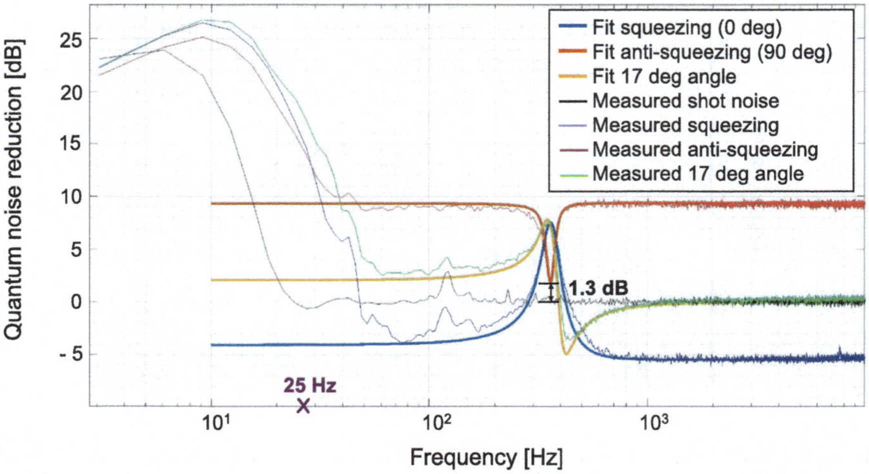

Fig. 5.30 First demonstration of squeezing ellipse rotation on FDS system in Virgo at 361 Hz. The black curve is the measured shot noise, in blue the measured and fitted anti-squeezing level, in red the measured and fitted squeezing level, in green/yellow the squeezed states detected at 17°

The first limitation is due to the scattered light produced by the local oscillator beam on the homodyne photodiodes, as we will describe later in Part III. The second problem is due to the low lock precision which increases the phase noise in the ellipse and marginally to the RTL of the cavity.

5.6.1 Lock Precision

In order to improve lock precision, the bandwidth of the longitudinal cavity loop must be increased. Using only FCEM as an actuator, it is not possible to go above 100 Hz of UGF. We therefore decided to send the correction also on the frequency of the squeezer's main laser, by acting on the frequency of the PLL between Virgo and the main laser. The conceptual diagram of the loop is shown in Fig. 5.31. Thus, a bandwidth of 100 Hz was measured for the part acting on the cavity mirror and $\sim$ 1 kHz for the part acting on the laser frequency. The UGF of the filter cavity length control is calculated to be $\sim$1 kHz. In this way a lock precision of 1 Hz has been achieved.

This allow to have the detuning of the filter cavity more stable in frequency and to have a deeper and less broad 'hole' of the anti-squeezing at the detuning frequency. This can be seen as an improvement of the phase noise losses.

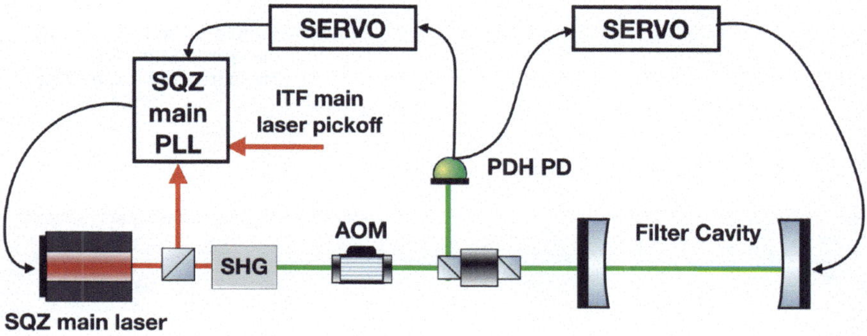

Fig. 5.31 Optical scheme of the filter cavity longitudinal control loop acting on both the end mirror and the main squeezer frequency

5.7 FDS Measurements at 25 Hz on EQB1

In summer 2022, FDS measurements on the EQB1 homodyne were repeated. The main improvements compared to the measurements carried out in autumn 2021 were:

- the reduction of propagation losses, having better aligned the double pass into the Faraday Isolators of the FDS on SQB1;
- the much improved lock precision thanks to the cavity lock loop shown in Fig. 5.31, leading to a phase noise reduction;
- the improved IR mode matching in the filter cavity, leading to an optical losses reduction;
- the fact that the measurements were carried out in summer where seismic noise is lower and therefore the bump noise due to scattered light is at lower frequencies;
- the fact that the automatic cavity control is done with wavefront sensing instead of dither lines on the cavity mirrors (more details in Part III).

Figure 5.32 shows the rotation of anti-squeezing at 25 Hz. The bump due to the scattered light does not allow one to see properly the red curve go below the shot noise (see Chap. 8), but according to the fit one should measure 2 dB of quantum noise reduction at low frequencies. The estimated optical losses are 14% : 7% are propagation losses, 3–4% due to HD contrast.

The comparison between the expected and measured losses for FDS is shown in Table 5.3

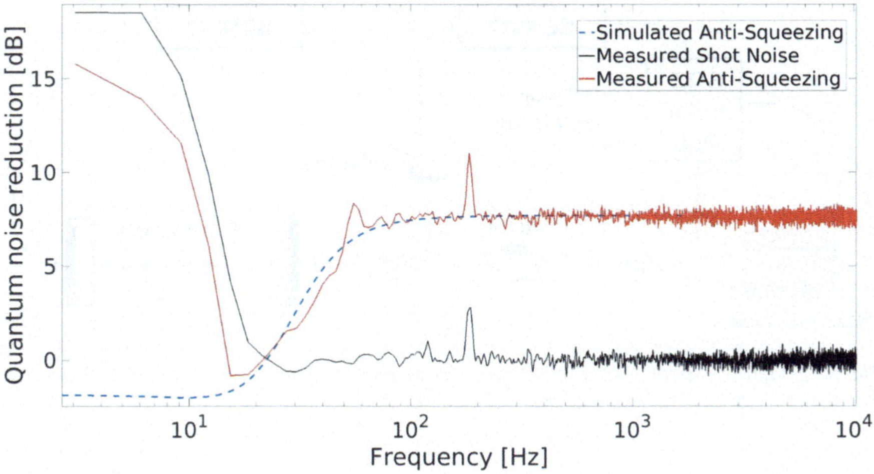

Fig. 5.32 FDS measurement with rotation frequency of 25 Hz. The black curve represents the shot noise and the red curve the anti-squeezing. The dashed blue curve is the fit of the data, showing that we should have around 2 dB of quantum noise reduction at low frequencies

Table 5.3 Comparison between the expected (with 4.5 dB quantum noise improvement at high frequency and 40 W of ITF input power) and measured (on EQB1 HD during the commissioning of the system) losses for FDS. (*) The measured phase noise doesn't take into account the one coming from the propagation in the ITF. (**) A FC length fluctuation of 8 Hz rms allows to measure 4.5 dB of high frequency quantum noise reduction in the ITF without spoiling the low frequency

Degradation source	Expected for O4	Measured in November 2021	Measured in June 2022
Filter cavity RTL	40 ppm	31 ppm	31 ppm
Phase noise*	20–60 mrad	30 mrad	30 mrad
FC rms length fluctuation	< 1 Hz	8.3** Hz	~1 Hz
FC mode mismatch	2%	1%	1%
SC to OPO mode mismatch	2%	1%	1%
Propagation losses	2–3%	13–15%	7%

5.8 Conclusion

In this chapter, we have described the FDS system installation and its commissioning. We described the installation of the optics on the suspended benches (SQB1 and SQB2) interfacing the squeezed state source with the filter cavity and the interferometer, first at LAPP and subsequently on the Virgo site. We then described how these suspended benches are controlled, both with the usual LVDT control system and with the PSDs, a new method we are implementing. Then, once the installation was complete, the beams were aligned towards the FDS system for the first time. Once the system was aligned and properly controlled, we were then able to carry out several campaigns of squeezing measurements with the homodyne detector. We started by measuring and characterizing the FIS states and then measuring the rotation of the squeezing ellipse. The first demonstration of the rotation of the ellipse took place at 361 Hz, but without reaching the shot noise level. This was due to both the low lock precision of the filter cavity and the light scattered by the homodyne photodiodes. After improving the lock precision and performing the measurements in a better environmental condition, we were able to obtain an FDS measurement at 25 Hz, 2 dB below the shot noise level. The next steps are to reduce the scattered light noise at low frequency to be able to see the squeezing rotation at low frequency. One strategy to solve this issue is the implementation of an active control loop, as will be described in Part III. Before injecting squeezed states into the ITF, the squeezed beam must be matched to the ITF beam. This will be described in Part IV.

Part III
Scattered Light Noise in Advanced Virgo Plus

Chapter 6
Stray Light in Gravitational Waves Detectors

Stray light is one of the major limitations of the sensitivity of interferometric gravitational wave detectors, as introduced in Sect. 2.2.6. The laser beam propagates through several optical elements that can back-reflect and scatter it. This light can recouple with the interferometer's main beam, adding a spurious field to the main one which degrades the detector performances [86–88].

To describe the coupling mechanism, one introduces the field E_{in} inside the interferometer at a reference point, the field E_{out} which is exiting from one of the ports of the detector and E_r the back-scattered field that re-couples with the interferometer beam at the same point where E_{in} is calculated. These fields are shown in Fig. 6.1.

The phase $\phi_r(t)$ of the field E_r encodes the relative (not constant) motion $z_r(t)$ between the interferometer and the scattering object, defined as

$$\phi_r(t) = \frac{4\pi}{\lambda} \cdot z_r(t). \tag{6.1}$$

The stray light field sums with the interferometer beam E_{in} as described in [89]

$$\begin{aligned} E_{in} + E_r = E_{in} + \sqrt{f_r} \cdot E_{out} \cdot e^{i(\phi_0 + \phi_r(t))} = \\ = E_{in} \cdot \left[1 + \sqrt{f_r \cdot \frac{P_{out}}{P_{in}}} \cdot e^{i(\phi_0 + \phi_r(t))} \right] \end{aligned} \tag{6.2}$$

where ϕ_0 is a static phase which depends on the scatterer position in the interferometer, f_r is the fraction of stray light power exiting the detector output port which is recoupled back into the main mode. It is considered that $f_r \frac{P_{out}}{P_{in}} << 1$. From

E. Polini, *Broadband Quantum Noise Reduction in Advanced Virgo Plus*,
Springer Theses, https://doi.org/10.1007/978-3-031-95143-5_6

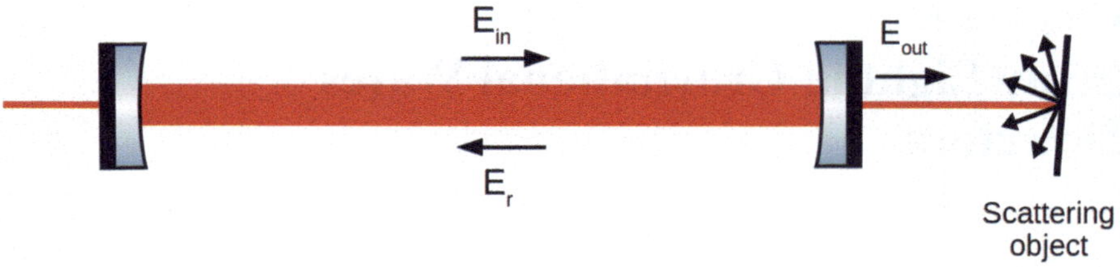

Fig. 6.1 Scattering process in the interferometer

Eq. 6.2 one can deduce that the effect of the scattered light noise is therefore given by amplitude and phase noise contributions, respectively described as

$$\frac{\delta P(t)}{P} = \sqrt{f_r \cdot \frac{P_{out}}{P_{in}}} \cdot \cos(\phi_0 + \phi_r(t)) = \sqrt{f_r} \cdot \frac{P_{out}}{P_{in}} \cdot \cos\left[\frac{4\pi}{\lambda}(z_0 + z_r(t))\right] \tag{6.3}$$

$$n_\phi(t) = \sqrt{f_r \cdot \frac{P_{out}}{P_{in}}} \cdot \sin(\phi_0 + \phi_r(t)) = \sqrt{f_r} \cdot \frac{P_{out}}{P_{in}} \cdot \sin\left[\frac{4\pi}{\lambda}(z_0 + z_r(t))\right]. \tag{6.4}$$

These noises encode the coupling of the scattered light with the interferometer and they can be propagated to the detector sensitivity using the linear response of the system, as [89]

$$\begin{aligned} h_r(f) = & \sqrt{f_r \cdot \frac{P_{out}}{P_{in}}} \cdot \left\{K_{n_\phi}(f) \cdot \mathcal{F}\left[\sin\left(\frac{4\pi}{\lambda}(z_0 + z_r(t))\right)\right]\right\} + \\ & + \sqrt{f_r \cdot \frac{P_{out}}{P_{in}}} \cdot \left\{K_{\frac{\delta P}{P}}(f) \cdot \mathcal{F}\left[\cos\left(\frac{4\pi}{\lambda}(z_0 + z_r(t))\right)\right]\right\} \end{aligned} \tag{6.5}$$

where $\mathcal{F}$ represents the Fourier Transform, $K_{n_\phi}(f)$ and $K_{\delta P/P}(f)$ are the phase and amplitude quadrature transfer functions (TFs) from the port where the scatterer is located to the GW signal, i.e. the linear response of the interferometer phase and amplitude noises. The noise on the sensitivity depends therefore on $\sqrt{f_r}$, $z_r(t)$ and these TFs.

The factor f_r receives several contributions

$$f_r = f_{sc} + f_{sp} + f_{Rayleigh} + f_{extra} \tag{6.6}$$

where f_{sc} accounts for the light scattered by the surface, f_{sp} for the specular reflection, $f_{Rayleigh}$ represents the light scattered by atoms or molecules (in crystals for example), f_{extra} is the extra contribution of scattered light coming from spurious beams. Figure 6.2 illustrates the specular reflection, given by the factor f_{sp} and the scattering contributions, given by the other terms.

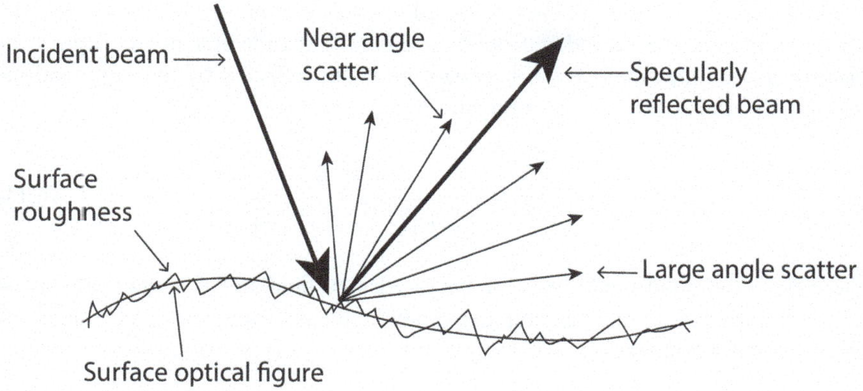

Fig. 6.2 Diagram of scattering produced by an optical surface roughness. The incident beam is either scattered in all directions, at large or small angles, or specularly reflected [90]

The factor f_{sc} is an intrinsic contribution of each optical element and is related to the bi-directional reflectance distribution function (BRDF) of the surface by means of the simplified relationship [89], demonstrated in [91],

$$f_{sc} \sim \frac{\mathrm{BRDF}(\theta) \cdot \lambda^2}{\pi \omega^2} \tag{6.7}$$

where ω is the beam radius at the scattering surface and θ the scattering angle. The integral of the BRDF gives the Total Integrated Scatter (TIS):

$$\mathrm{TIS} = \int_\theta \mathrm{BRDF}(\theta) 2\pi \cos\theta \sin\theta d\theta. \tag{6.8}$$

For a rough surface with defects, the TIS can be calculated from

$$TIS = \left(\frac{4\pi\sigma_{RMS}}{\lambda}\right)^2 \tag{6.9}$$

where σ_{RMS} is the surface micro-roughness. If the surface scatters uniformly in all directions, then a good approximation of the BRDF is

$$\mathrm{BRDF}(\theta) = \frac{TIS}{\pi}. \tag{6.10}$$

The contribution of f_{sc} cannot be eliminated but it can be reduced by replacing components that back scatter less light. Since it is not easy to make numerical simulations of this quantity, the strategy used is to directly measure this parameter experimentally as will be described in Sect. 6.2.

The factor f_{sp} is the light back reflected towards the interferometer by specular reflection in case the considered optic is almost perpendicular to the beam. The quantity that defines the overlap between the beam reflected by the optic and the incoming beam is called Overlap Integral (OI), given by

$$OI = |\langle E_2|E_1\rangle|^2 = \frac{\left|\int E_1^* E_2 dA\right|^2}{\int |E_1|^2 dA \int |E_2|^2 dA} \tag{6.11}$$

where E_1 is the input field, E_2 is the back-reflected field and A is the beam cross-section. Equation 6.11 can be computed analytically assuming the input beam to be a pure Gaussian beam [89]

$$f_{sp} = \alpha\,|\langle E_2|E_1\rangle|^2 = \frac{\alpha R^2 z_R^2 \exp\left[-\frac{2\pi D^2 z_R \beta^2}{2\lambda}\left(\frac{1}{D^2+z_R^2}+\frac{1}{(D-R)^2+z_R^2}\right)\right]}{(D^2+z_R^2)\left[(D-R)^2+z_R^2\right]} \tag{6.12}$$

where α is the reflectivity of the optics, R is the radius of curvature of the reflecting surface, D is the distance between the surface and the beam waist position, z_R is the beam Rayleigh range and β is the small angle at which the optic is tilted. These parameters are illustrated in Fig. 6.3. The elements that reflect the most are the lenses. Since this amount decreases exponentially as a function of the angle β at which the elements are tilted, the lenses have almost all been tilted by at least a hundred μrads to make this contribution negligible.

The factor $f_{Rayleigh}$ arises from the scattering of light by particles much smaller than the wavelength of the radiation, like in crystals. In Virgo's optical systems, this

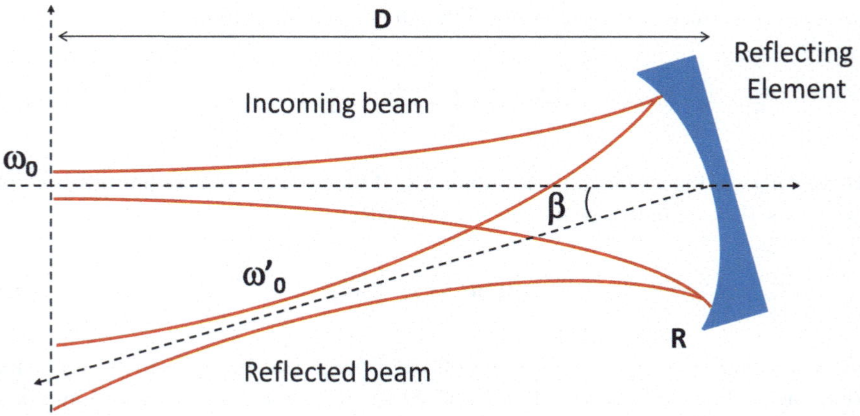

Fig. 6.3 A Gaussian beam arrives on a reflective surface and it is reflected with an angle β [89]

type of scattering occurs from the Terbium Gallium Garnet (TGG) crystals within the Faraday Isolators and the output mode cleaner cavity (Suprasil 3001).

The factor f_{extra} refers to all the other contributions. Actually, in Virgo's optical systems, spurious beams can be generated and then they can be reflected by vacuum chambers, optical mounts, optics or other elements, generating scattered light. This propagates through the system and can re-couple with the interferometer beam. By tracing these beams, called ghost beams, it is possible to mitigate them before the generation of scattered light (see Chap. 7).

6.1 Scattered Light from Suspended Benches in O3

The relative motion of the scattering optics $z_r(t)$ is related to the relative displacement between the suspended benches on which the optics are installed and the interferometer, introduced by the seismic activity.

From Eq. 6.5, it can be noticed that

- for small displacements of the scattering object, $z_r(t) << \frac{\lambda}{4\pi}$, $h_r(t)$ depends linearly on $z_r(t)$;
- for a large displacement of the scattering object, $z_r(t) > \frac{\lambda}{4\pi}$, the coupling to $h_r(t)$ is non-linear and the so-called *fringe wrapping* occurs. In particular, if the scattering object moves with a sinusoidal motion of amplitude $> \frac{\lambda}{4\pi}$, the spectra $\mathcal{F}[\sin(\phi_0 + \phi_r(t))]$ and $\mathcal{F}[\cos(\phi_0 + \phi_r(t))]$ of Eq. 6.5 become of same amplitude and with a *shoulder* shape (as illustrated in Fig. 6.4) with a cut-off at [92]

$$f_{max} = \frac{2v_{sc}}{\lambda} \tag{6.13}$$

where v_{sc} is the velocity amplitude of the scatterer. In this regime, the strain $h_r(f)$ can be approximated as [93]

$$h_r(f) \propto \sqrt{\frac{f_r}{f_{max}}} \cdot TF(f) \quad \text{for} \quad f < f_{max}. \tag{6.14}$$

During O3, measurements were made to estimate the effect of scattered light from suspended benches on Virgo's sensitivity curve. To do this, suspended benches were shaken to artificially increase $z_r(t)$, resulting in the curves shown in Fig. 6.4: in red the reference sensitivity of the ITF, i.e. without injecting noise, in blue h(t) during the noise injection and in yellow the reference h(t) plus the scattered light projection using Eq. 6.5.

From the measurements and Eq. 6.14 it was possible to estimate of the value f_r for each bench, as reported in Table 6.1. SNEB and SWEB are symmetrical and therefore have the same optical configuration but the results show a different f_r

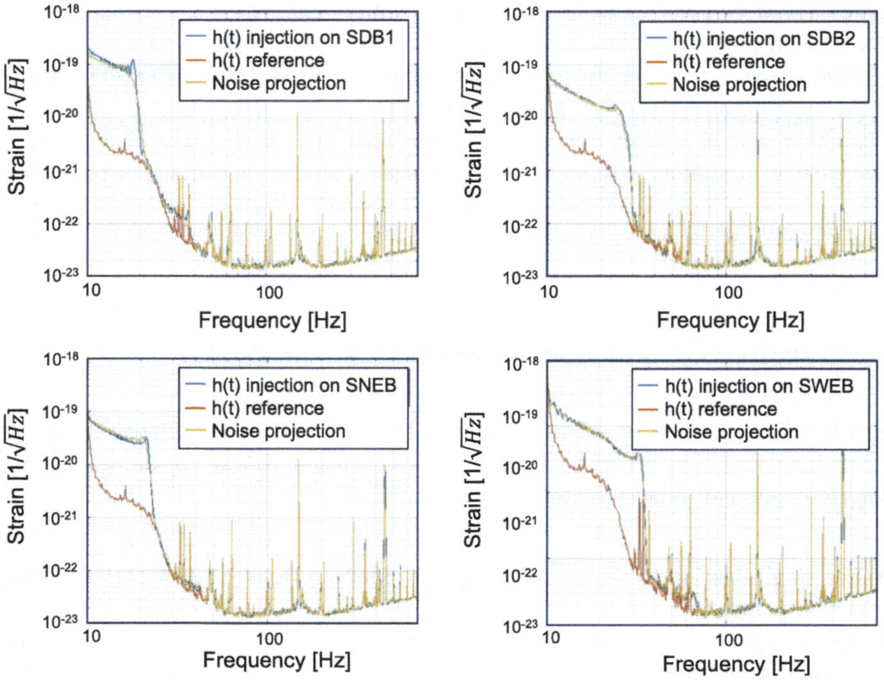

Fig. 6.4 Stray light projection fitted to the measured coupling during noise injection on different suspended benches (SDB1, SDB2, SNEB, SWEB). The red curve represents the reference sensitivity of the ITF, i.e. without injecting noise The blue curve represents h(t) during the noise injection and the yellow one the reference h(t) plus the fit of the stray light projection

Table 6.1 Fraction of re-coupled scattered light f_r from the suspended benches measured in O3

Bench	f_r measured in O3
SDB1	$2 \cdot 10^{-9}$
SDB2	$4 \cdot 10^{-10}$
SNEB	$3 \cdot 10^{-8}$
SWEB	$5 \cdot 10^{-9}$

value. The reason is that on SNEB the reflections of quadrant photodiodes were not properly dumped and so there was much more scattered light on this bench.

These values of f_r can be compared with the expectation from the scattering of all the optical elements of these benches using Eqs. 6.7 and 6.12. From this computation it comes out which are the most critical elements on each bench. This is useful in order to mitigate the noise coupled with the interferometer. Using these estimates we simulate the projections of the scattered light noise expected for O4 to see if it limits the sensitivity. In order to get an experimental estimate of the scattered light produced by the different optical elements on the suspended benches, we developed a back-scatterometer in the laboratory at LAPP that can measure the f_{sc} factor directly. This is presented in the next section.

6.2 High-Angular Resolution Back-Scattermeter

In order to measure the amount of light backscattered from an optic and that re-couples with the main beam, we built an interferometric back-scattermeter at LAPP. The experimental setup is shown in Fig. 6.5. A laser beam at $\lambda = 1064$ nm propagates towards a 50/50 beam splitter where the transmitted part will be used as a local oscillator (LO) for homodyne detection (see Sect. 3.3). The reflected beam, on the other hand, propagates up to the scattering sample of which one wants to measure f_{sc} i.e. the BRDF. The dimension of this beam is amplified from a waist of 300 μrad to 2.1 mm using a telescope composed of three lenses. The sample is free to move along the direction of the beam, as it is mounted on a small bench suspended with four metal wires. The free motion of this suspension creates a sinusoidal motion $\Delta L(t)$ of few hundred wavelengths (fringe wrapping case). The sample is tilted to an angle at which we want to measure the BRDF. The scattered light is reflected back to the homodyne where it couples with the LO. The LO field can be written as

$$\mathbf{E}_{LO} = E_{LO} \cdot e^{i\phi_{LO}} \tag{6.15}$$

and from Eq. 6.2, the scattered light field can be written as

$$\mathbf{E}_{sc} = \frac{1}{\sqrt{2}} \cdot E_{LO} \cdot \sqrt{f_{sc}} \cdot e^{i\phi_r(t)} \tag{6.16}$$

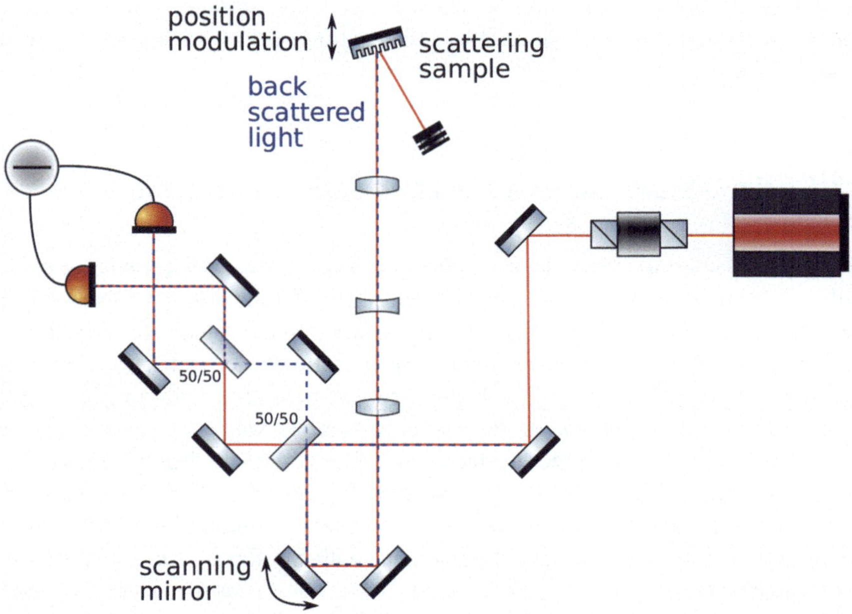

Fig. 6.5 Back-scatter meter experiment set up at LAPP [91]. The back-scattered light is represented by the dashed blue line

where the factor $\frac{1}{\sqrt{2}}$ comes from the splitting of the scattered light field, as shown in Fig. 6.5. The two photodiodes of the homodyne reveal the powers associated with the fields described in Eqs. 3.48 and 3.49 in Sect. 3.3. Calling $P_1(t)$ and $P_2(t)$ the powers detected by each photodiode, the differential channel is:

$$\begin{aligned} P_1(t) - P_2(t) &= \frac{1}{2}\left|\mathbf{E}_{LO} + \mathbf{E}_{sc}\right|^2 - \frac{1}{2}\left|\mathbf{E}_{LO} - \mathbf{E}_{sc}\right|^2 = \\ &= 2 \cdot \mathcal{R}(\mathbf{E}_{LO} \cdot \mathbf{E}_{sc}) = P_{LO} \cdot \sqrt{2 \cdot f_{sc}} \cdot \cos(\phi_r(t)) \end{aligned} \tag{6.17}$$

where the static phase of the LO has been omitted. This signal can be normalized by the total power received by the photodiodes $P_1(t) + P_2(t) \sim P_{LO}$ obtaining [91]:

$$A(t) = \frac{P_1(t) - P_2(t)}{P_1(t) + P_2(t)} = \sqrt{2 \cdot f_{sc}} \cos\left(2\pi \frac{\Delta L(t)}{\lambda}\right) \tag{6.18}$$

where $\frac{2\pi}{\lambda}\Delta L(t) = \phi_r(t)$, with $\Delta L(t)$ being the position modulation of the scattering sample. For $\Delta L >> \lambda$ the factor f_{sc} associated to the scatterer can therefore be computed as the variance of the quantity $A(t)$ defined in Eq. 6.18.

The system was aligned positioning a mirror at the location of the scattering sample, obtaining a contrast of 88%. The power spectral density of the differential homodyne signal from a scattering object is spread between 0 Hz and a few kHz. This measurement is contaminated by the speckle, which is the particular random realization of the back-scattered light. This random effect was mediated by scanning the scattering sample in translation. We scan the target surface with a motorized mirror, spanning ±3 beam radii and we assumed the surface to be uniform in the area of the scan.

6.2.1 Back-Scattering as a Function of the Beam Radius

We used this setup to verify the relationship in Eq. 6.7, i.e. the dependence of f_{sc} with the beam size at the scattering surface $w(z)$. To do this, we measured the backscattering from a PTFE target, i.e. integrating sphere plug, as a function of the beam size. PTFE is close to being a perfect scatterer, i.e. scattering light in all directions with a BRDF $= \frac{1}{\pi}$. The beam size was measured with a CCD camera at the different positions where the scattering sample was subsequently placed. Figure 6.6 shows the measured f_{sc} factor as a function of the beam size. The red curve represents the theoretical curve expected from Eq. 6.7. The blue data shows the measured f_{sc} as a function of the beam size with a configuration of four lenses. It can be seen that the data are in agreement with the theory for a beam radius greater than 200 μm while they are a factor 2–3 below it near the waist. This could be due to the aberration in the beam converged using a single plano-convex lens. In black the data for the configuration with a single lens of 175 mm focal length

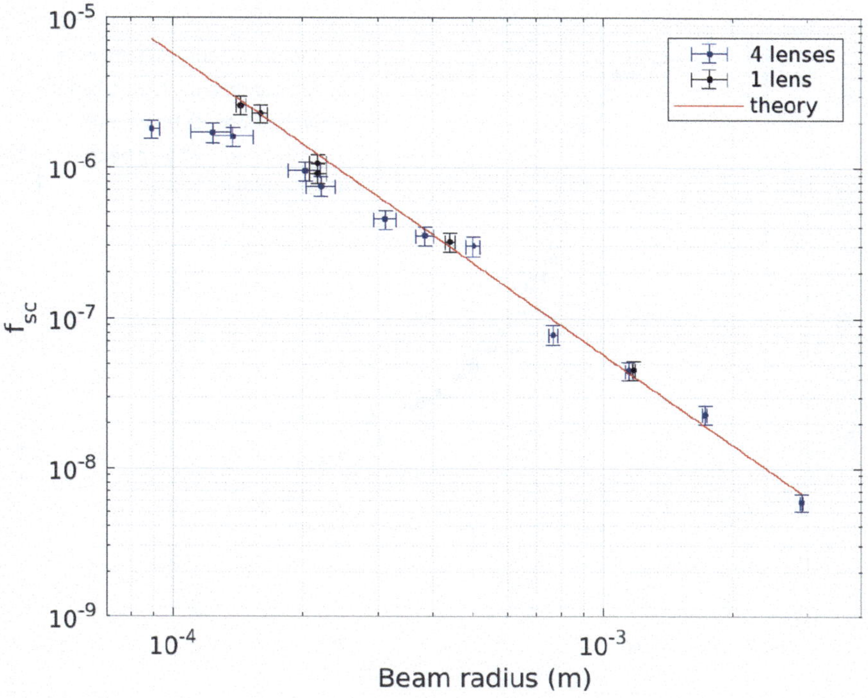

Fig. 6.6 Measured back-scattered light fraction f_{sc} as a function of the beam size: for a PTFE sample: in red the theoretical behavior described in Eq. 6.7, in blue the data for a four-lenses configuration and in black the data for a single lens configuration. The vertical error bars correspond to the statistical error of the measurement due to speckle averaging, while the horizontal error bars correspond to the target positioning errors of ±5 mm re-stated as errors on the beam radius on the target [91]

and shorter optical path are shown. In this case the aberration effect is reduced and there is an agreement with the theory for all the radii down to a waist of 145 μm. In conclusion, the experimental results show agreement with Eq. 6.7 and showing that the factor f_{sc} is inversely proportional to the square of the beam size at the scattering object.

6.2.2 Back-Scattering as a Function of the Incident Angle

Then we used our setup to measure the scattering as a function of the incidence angle. The beam is sent on the scatterer with an angle of incidence θ. In order to take this into account, the Eq. 6.7 can be re-written as

$$f_{sc} = \mathrm{BRDF}(\theta)\cos(\theta)\frac{\lambda^2}{\pi\omega^2}. \tag{6.19}$$

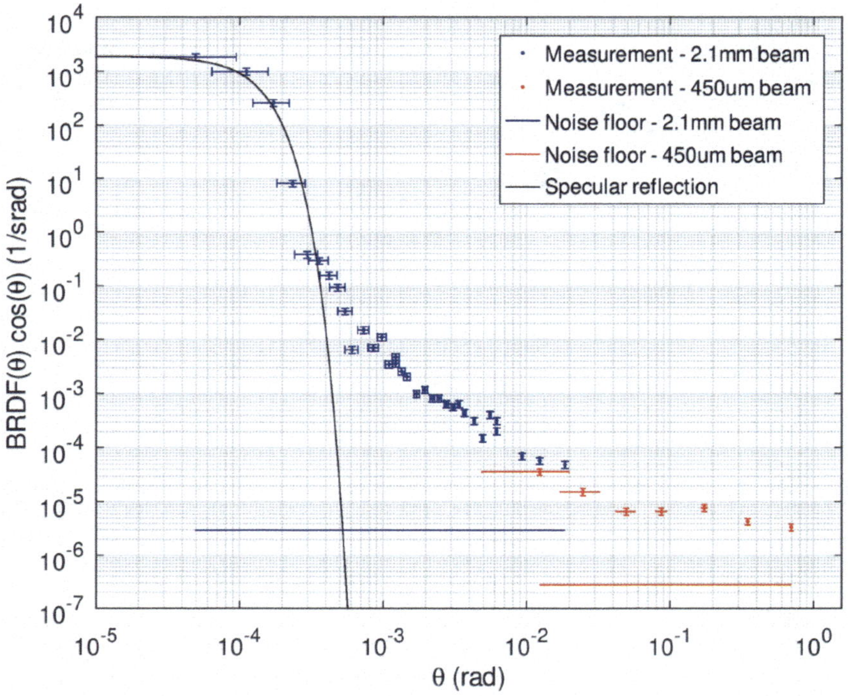

Fig. 6.7 Measured BRDF using a 2.1 mm radius beam (blue data points) and a 450 μm radius beam (red data points) as a function of the incidence angle θ. The horizontal blue and red lines represent the noise floors measured without a sample. The black line represents the expected specular reflection pattern (Eq. 6.12) from the window with 700 ppm reflectivity

We started with a multi-band anti-reflective coated vacuum window as a scattering sample, with a measured reflectivity of 700 ppm at 1064 nm for 0° of incidence angle. The sample was rotated by small angles horizontally and the f_{sc} factor was measured for each angle. This was translated into BRDF using Eq. 6.19 and the result for different angles is shown in Fig. 6.7. The specular reflection tail of the Gaussian beam is dominant for $\theta < 500$ μrad as the beam leaving the telescope has a divergence of 160 μrad. The angular errors are due to the sum in quadrature of the ± 30 μrad beam incidence angle variation during the sample surface scan and of the ± 30 μrad positioning error of the kinematic mount micrometer. For angles between 1 mrad and 10 mrad, the BRDF is proportional to $\frac{1}{\theta^2}$, which is typical of polished optics [89]. However, for angles greater than 50 mrad, the BRDF becomes independent of the incidence angle. The setup at LAPP allows to measure the light back-scattered from optical components with a 160 μrad angular resolution and incidence angle larger than 500 μrad [91]. In particular, it is able to measure back-scatter at angles of few mrad that are relevant for beam expanding telescope lenses in gravitational wave detectors. With this setup it is possible to measure the scattering of coated optics, which depends on surface defects and roughness, but also on

Table 6.2 Values of the $BRDF(\theta)$, where θ is the incidence angle, measured at LAPP with the back-scattermeter described in Sect. 6.2.1

Element	θ	$BRDF(\theta)$
SDB1 quadrants (B5)	5°	$3.5 \cdot 10^{-3}$
SDB2/SNEB/SWEB/SPRB quadrants	5°	3
SDB2/SNEB/SWEB/SPRB photodiodes	20°	$4.9 \cdot 10^{-5}$
SNEB/SWEB Silicon beam dump	1°	$1.0 \cdot 10^{-4}$

the anti-reflective coating. We characterized the back-scattered light produced by several elements of the Virgo benches (photodiodes, quadrant photodiodes, beam dumps), as reported in Table 6.2. These values will be used in Sect. 6.3.

6.3 Scattered Light Projections for O4

It is possible to simulate the projections of the scattered light noise arising from the suspended benches for O4 using Eq. 6.5. This work has been performed for the benches SDB1, SDB2, SNEB, SWEB and SPRB, defined in Table 2.2. In the following, the method used to compute the scattered light noise projections associated to the suspended benches is described, see Sect. 6.3.1. Moreover, it will be described the procedure used to estimate the factors f_r coming from each suspended bench in O4, see Sect. 6.3.2. Those values are compared with the measurements and estimate of O3. Finally the noise projections for O4 are shown in Sect. 6.3.7.

6.3.1 Method

The ingredients of Eq. 6.5 are the transfer functions of the interferometer, the relative displacement between the detector and the scatterer $z_r(t)$ and the factor f_r.

The ITF transfer functions, $K_{n_\phi}(f)$ and $K_{\frac{\delta P}{P}}(f)$, depend on the detector optical configuration and where the scatterer is located within the ITF. They can be computed with Optickle [94], a Matlab based software used to generally model the electro-opto-mechanical response of a GW detector. On the top of Fig. 6.8, it is shown the example of the transfer functions for the bench SDB1, $K_{n_\phi}(f)$ in blue and $K_{\frac{\delta P}{P}}(f)$ in red using the ITF configuration of O4.

The relative non-stationary displacement between the intererometer and the considered suspended bench $z_r(t)$ can be measured by sensors placed on the superattenuators of the ITF mirrors and on the bench suspensions. The effect of scattered light on the sensitivity curve of the detector is bigger with high microseism since the cut-off frequency of the scattered light shoulder is proportional to the velocity of the scattering object, as shown in Eq. 6.13. If the microseism is high, the velocity of the suspended bench is greater and thus the low-frequency region

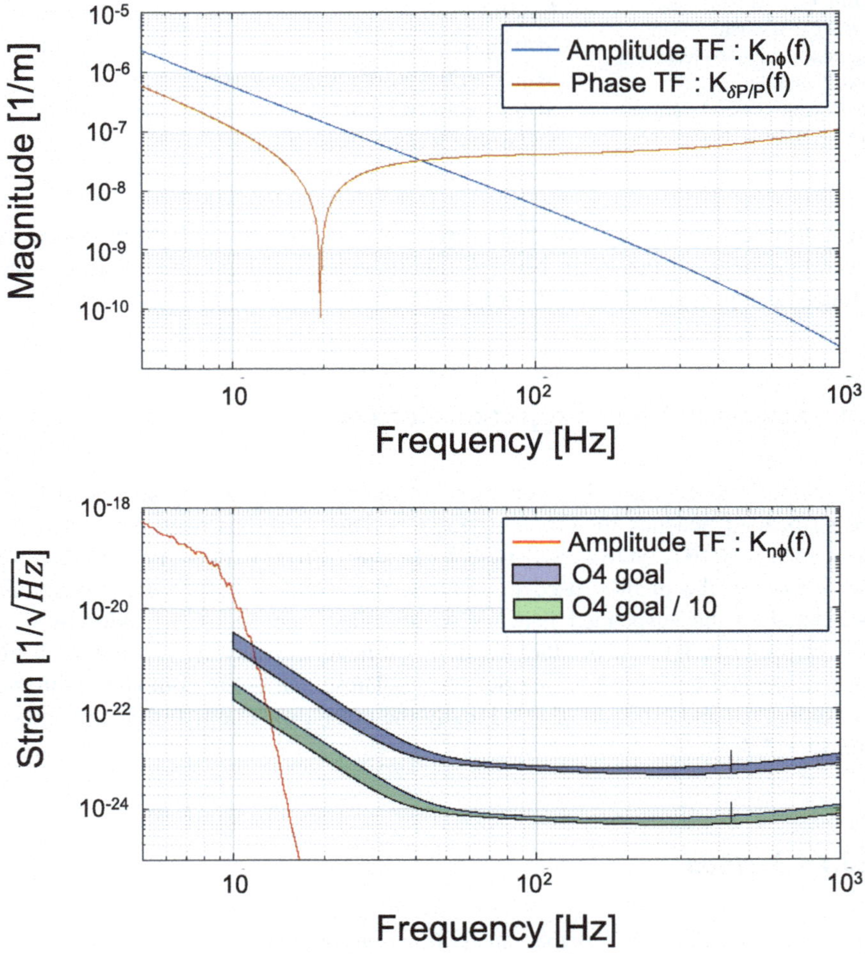

Fig. 6.8 On the upper plot, the SDB1 transfer functions $K_{n_\phi}(f)$ and $K_{\delta P/P}(f)$ are shown for O4. On the bottom plot, the scattered light noise projections for SDB1 are shown using the f_r measured in O3 and scatterer relative motion shown in Fig. 6.9

worsened by the scattered light noise is larger. We chose to consider the data in a high-seismic period during the O4 commissioning, verifying the signals of the seismometers located near the towers under investigation. In Fig. 6.9, the signals of the seismometers in the central building (CEB), which houses SDB1 and SDB2 benches, and in the terminal buildings (NEB/WEB), which house SNEB and SWEB, are shown for the time chosen for the projections. The seismic motion at the chosen time is about 5–10 times higher than during quiet times.

The example of scattered noise projection from SDB1, using the f_r values measured during O3 (see Sect. 6.1), considering the transfer functions of O4 (on the

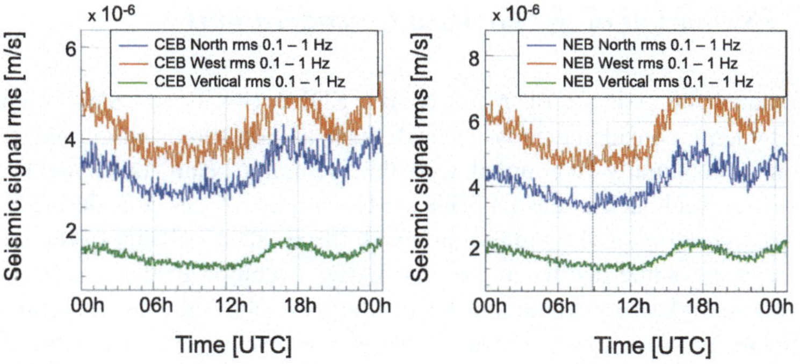

Fig. 6.9 Seismometers signals in the Central Building (CEB) and in the North terminal building (NEB). The rms of the signal between 0.1 and 1 Hz is shown, since it carries the information on the micro-seismic activity which couples with the bench motion. The different lines represent the three directions of the motion: North (N), West (W) and Vertical (V)

top of Fig. 6.8) and the $z_r(t)$ motion shown in Fig. 6.9, is presented on the bottom of Fig. 6.8. It can be seen that at low frequency the sensitivity curve could be spoiled by the scattered light noise coming from the suspended bench SDB1 if the f_r value is the same as in O3.

To better understand the limiting elements, we computed the expected f_r values considering all the individual optical elements on the benches. The factors f_{sp} have been calculated using 6.12 and the factors f_{sc} using Eq. 6.7 where the BRDF is measured with the back-scatterometer described in Sect. 6.2.1 or from literature. Since the optics on the bench do not directly receive all the incoming power on the bench, the f_r factors must be normalized by a factor

$$\alpha_p^2 = \left(\frac{P_{optic}}{P_0}\right)^2 \tag{6.20}$$

where P_{optic} is the power arriving on the scattering object and P_0 is the power at the interferometer port where the transfer function is computed. This factor takes into account all the splitting elements on the beam path and the square accounts for the fact that the beam passes twice through these elements, on the incoming and returning ways. The difference between the optical configurations between O3 and O4 lies in the replacement of some beam splitters along the optical path and thus the α_p values vary between O3 and O4. It is interesting to compare the f_r values measured during O3 with the values computed with the optical configuration of O3 to see if they are consistent. However, the interest for O4 is the estimation of f_r with the new configuration to predict the effect of scattered light during the next run. The estimation of f_r for the different benches are described in the next sections.

6.3.2 Estimation of f_r: General Considerations

We calculated the estimate of f_r for SDB1, SDB2, SNEB and SWEB, studying all the elements considered most critical, such as non-tilted lenses whose retro-reflection is maximally re-coupled with the incoming beam, or elements whose BRDF is high such as the sensors. This study was carried out considering both the optical configuration of O3 and O4, in which the α_p factor changes. The elements with a large f_{sc} value due to an elevated BRDF common to the benches studied are the photodiodes, quadrants and beam dumps in Silicon. The cameras have a large BRDF but since the fraction of power α_p reaching them is several order of magnitude smaller than the other elements, their contribution is negligible on all the benches. On SDB1, the TGG crystals inside the Faraday Isolator and the OMC cavity are found in addition. The Faraday Isolator on SDB1 attenuates by a factor $\alpha_{FI} = 10^4$ all the back-reflected and back-scattered light from the elements positioned after it, on the B1p optical path. This is the case for most of the elements on SDB1 and all the elements on SDB2. Concerning the lenses, where possible they were tilted during the commissioning of O4 in order to prevent their back reflection from re-coupling with the main beam. However, there are large lenses perpendicular to the incoming beam at the entrance of most of the considered benches that cannot be tilted due to mechanical constraints. These elements are found to contribute to the scattered light with a large f_{sp} factor, considering a null angle of incidence.

6.3.3 Estimation of f_r for SDB1

The O4 optical configuration of SDB1 is shown in Fig. 6.10.

As shown in Fig. 6.10, there are two separate and distinct beams arriving from the interferometer: B1p and B5. Since their transfer functions are different, the two optical paths must be considered separately. B1p passes through the Faraday Isolator which attenuates by a factor $\alpha_{FI} = 10^4$ any back-reflection generate after it that would return back to the ITF, while B5 bypasses it. The elements studied on SDB1 are:

- TGG crystal of the Faraday Isolator both the surface scattering and the Rayleigh scattering;
- OMC cavity both the surface scattering and the Rayleigh scattering;
- Meniscus lens, i.e. L1, the first big lens at the entrance, perpendicular to the incoming beam (for both B1p and B5 beams);
- Silicon beam dump on B1s optical path;
- B5 quadrants.

The f_{sc} contributions arising from these elements are computed using Eqs. 6.10 and 6.7 and they are reported in Table 6.3. The f_{sp} contributions from the two surfaces of the meniscus lens are computed using Eq. 6.12 and they are reported

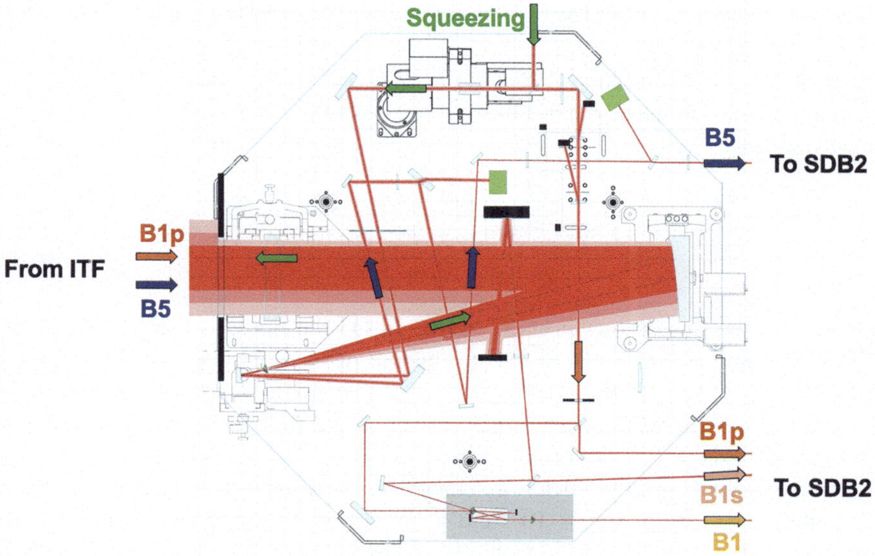

Fig. 6.10 Optocad scheme of SDB1 bench. On the top there is the Faraday Isolator containing the TGG crystal and on the bottom, in the grey box, there is the OMC cavity. A fraction of the power reflected by the OMC is dumped on a Silicon beam dump, indicated with the black square. A fraction of the B5 beam reaches two quadrants on SDB1, indicated by the green small boxes on the top right area of the bench

Table 6.3 Values of f_{sc} of the optical elements on SDB1 computed from Eq. 6.7

Element	α_p	$w(z)$	α_{FI}	$BRDF(\theta)$	f_{sc}
TGG surface	1	1.3 mm	no	$4 \cdot 10^{-6}$	$8.5 \cdot 10^{-13}$
TGG Rayleigh	1	1.3 mm	no	$3 \cdot 10^{-3}$	$3.1 \cdot 10^{-10}$
OMC surface	0.992	320 μm	yes	$4.4 \cdot 10^{-7}$ (O4) $1.8 \cdot 10^{-6}$ (O3)	$6.8 \cdot 10^{-11}$ (O4) $1.8 \cdot 10^{-11}$ (O3)
OMC Rayleigh	0.992	320 μm	yes	$5.6 \cdot 10^{-6}$	$8.6 \cdot 10^{-10}$ (O4) $2.5 \cdot 10^{-11}$ (O3)
B1s beam dump	0.92	2.8 mm	yes	$2.2 \cdot 10^{-5}$[a]	$9 \cdot 10^{-17}$
B5 QD1	0.085	1.0 mm	no	$3.5 \cdot 10^{-3}$	$9.1 \cdot 10^{-12}$
B5 QD2	0.188	0.6 mm	no	$3.5 \cdot 10^{-3}$	$1.2 \cdot 10^{-10}$

[a] The value of the TIS of this beam dump has been measured in 2010 at the Virgo site [95] and the BRDF has been obtained using Eq. 6.10

in Table 5.4. The beams and the optics characteristics used for the computations are reported in these tables as well.

The only difference between O3 and O4 consists in the replacement of the two low finesse output mode cleaner cavities (OMCs) with a high finesse one (see Chap. 10). This changes the amount of scattered light produced by the OMC since the back-scattered light is amplified by the cavity by the square of the number of round-trips N [96], defined as

Table 6.4 Values of f_{sp} of the meniscus lens L1 on SDB1 computed from Eq. 6.12, considering an incidence angle $\beta = 0$

Element	α (ppm)	RoC (m)	z_R (m)	D (m)	f_{sp}
L1 1st surface	190	1.072	$9.3 \cdot 10^{-3}$	-3.6	$6.8 \cdot 10^{-11}$
L1 2nd surface	42	3.195	$52.4 \cdot 10^{-3}$	10.3	$2.2 \cdot 10^{-10}$

$$N = \frac{F}{\pi} \tag{6.21}$$

where F is the finesse of the OMC cavity. In O3 the finesse of the cavities was ~125, meaning $N(O3)$~40. In O4 the finesse of the cavity is ~1000, meaning $N(O4)$~320. The BRDF coming from the internal surfaces of the cavity is computed from Eq. 6.9, considering a surface micro-roughness $\sigma_{RMS} = 0.1$ nm for O4 and $\sigma_{RMS} = 0.3$ nm for O3 [74].

On the other hand, the BRDF associated to the Rayleigh scattering has been computed fusing the TIS = $7 \cdot 10^{-7}$ cm^{-1} [97] of the Fused Silica, using Eq. 6.10 and multiplying it for the cavity geometrical length 25 cm. As reported in Table 6.3, the two contributions are comparable.

Similarly to the OMC cavity, f_r is calculated for the TGG crystal. The BRDF associated with the crystal surface scattering is calculated by Eq. 6.9, considering a surface micro-roughness $\sigma_{RMS} = 0.2$ nm [98]. The BRDF associated to the Rayleigh scattering has been computed from the TIS = $5 \cdot 10^{-4}$ cm^{-1} [99], using Eq. 6.10, multiplying it by the cavity length 18 cm and by a factor 0.5 due to a partial rotation of the polarization in the TGG crystal. As reported in Table 6.3, the main contribution is given by the Rayleigh scattering of the TGG crystal.

Along the path of B1p/B1/B1s, the contributions coming from the TGG crystal, the OMC cavity and the meniscus lens are of the same order of magnitude and can be summed up to obtain a factor $f_r \sim 1.5 \cdot 10^{-9}$ for O4 and $f_r \sim 6.4 \cdot 10^{-10}$ for O3. These values are close to the value measured in O3 (see Table 6.1). The fact that the O3 simulated value is lower than the measured one may be attributed to neglected effects such as dust contamination, which can increase the BRDFs. The next step will be to measure in the laboratory the BRDF of the TGG crystal and the OMC cavity with the set up described in Sect. 6.2, in order to have a more accurate value. Moreover, in order to reduce the contribution of the meniscus lens, it has been considered to replace its mechanical support so that it can be remotely rotated to reduce the scattered light noise it might introduce.

Along the path B5, the three lenses on the B5 path have been tilted during the O4 commissioning to make their contribution to f_{sp} negligible. The quadrants are tilted and the reflections have been properly stopped. In conclusion, the meniscus lens and the quadrant QD1 contributions are of the same order of magnitude and the overall contribution is $f_r \sim 4.1 \cdot 10^{-10}$, for both O3 and O4.

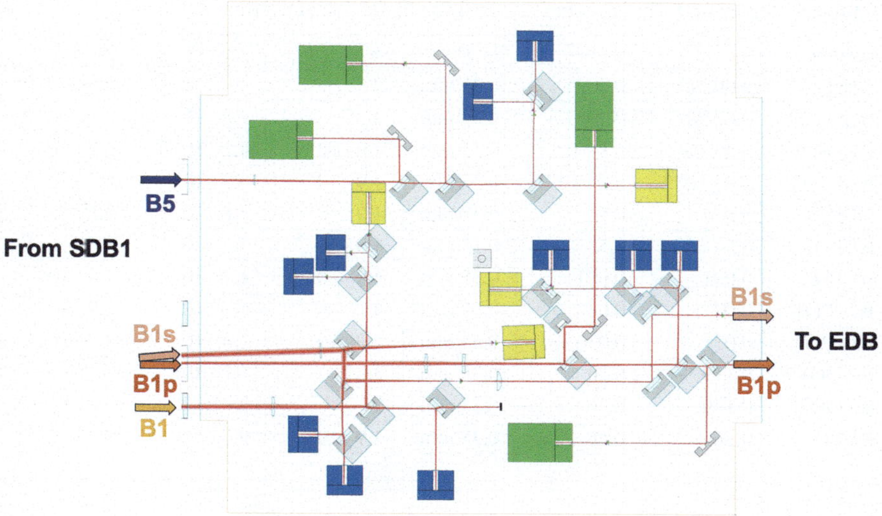

Fig. 6.11 Optocad scheme of SDB2 bench. The blue boxes represent the photodiodes, the green boxes represent the quadrants and the yellow boxes represent the cameras

6.3.4 Estimation of f_r for SDB2

The optical design of the SDB2 bench for O4 is shown in Fig. 6.11. There are several paths of different beams coming from SDB1 but there are only two re-coupling mechanisms: the B1/B1p and B1s beams that re-couple with the main interferometer beam and the B5 beam through the BS Anti-Reflective (AR) face. The power ratio α_p is computed with respect to the entrance of SDB1 since also the SDB2 transfer function is referred to the detector dark fringe port. Between O3 and O4, several beam splitters have been replaced on this bench.

The elements studied on SDB2 are:

- B1p quadrants QD1 and QD2;
- B1p photodiodes PD1 and PD2;
- B1 photodiodes PD1, PD2 and PD3;
- B1s photodiodes PD1 and PD2;
- B5 quadrants QD1 and QD2;
- B5 photodiode PD2.

Photodiodes measuring the p-polarization have not been considered as the re-coupling to the s-polarized main beam is assumed to be negligible .

Since this bench is located after the FI on SDB1, all the back-scattered light along the B1p/B1/B1s optical paths is attenuated by the factor $\alpha_{FI} = 10^4$, while the scattering produced by the optics along the B5 path is not attenuated. The BRDFs of these elements has been measured at LAPP and their values are listed in Table 6.2. The lenses on the different optical paths have been suitably tilted during

Table 6.5 Values of f_{sc} of the optical elements on SDB2 computed from Eq. 6.7

Element	α_p O3	α_p O4	$w(z)$	$BRDF(\theta)$	f_{sc} O3	f_{sc} O4
B1p QD1	0.00298	0.0025	0.35 mm	3	$7.8 \cdot 10^{-15}$	$5.5 \cdot 10^{-15}$
B1p QD2	0.00298	0.00375	0.35 mm	3	$7.8 \cdot 10^{-15}$	$1.2 \cdot 10^{-14}$
B1p PD1	$5.4 \cdot 10^{-3}$	$6.8 \cdot 10^{-4}$	0.35 mm	$4.9 \cdot 10^{-5}$	$4.2 \cdot 10^{-19}$	$6.7 \cdot 10^{-21}$
B1p PD2	$4.8 \cdot 10^{-4}$	$6.0 \cdot 10^{-5}$	0.35 mm	$4.9 \cdot 10^{-5}$	$3.3 \cdot 10^{-21}$	$5.2 \cdot 10^{-23}$
B1 PD1	0.5	0.5	0.35 mm	$4.9 \cdot 10^{-5}$	$3.6 \cdot 10^{-15}$	$3.6 \cdot 10^{-15}$
B1 PD2	0.5	0.5	0.35 mm	$4.9 \cdot 10^{-5}$	$3.6 \cdot 10^{-15}$	$3.6 \cdot 10^{-15}$
B1 PD3	0.001	0.001	0.35 mm	$4.9 \cdot 10^{-5}$	$1.4 \cdot 10^{-20}$	$1.4 \cdot 10^{-20}$
B1s PD1	0.07	0.036	0.35 mm	$4.9 \cdot 10^{-5}$	$1.4 \cdot 10^{-16}$	$1.9 \cdot 10^{-17}$
B1s PD2	0.07	0.036	0.35 mm	$4.9 \cdot 10^{-5}$	$1.4 \cdot 10^{-16}$	$1.7 \cdot 10^{-17}$
B5 QD1	0.119	0.26	0.35 mm	3	$1.2 \cdot 10^{-7}$	$6 \cdot 10^{-7}$
B5 QD2	0.149	0.26	0.35 mm	3	$2 \cdot 10^{-7}$	$6 \cdot 10^{-7}$
B5 PD2	0.26	0.26	0.35 mm	$4.9 \cdot 10^{-5}$	$9.7 \cdot 10^{-12}$	$9.7 \cdot 10^{-12}$

the commissioning for O4 to make their contribution to f_{sp} negligible. In Table 6.5, the f_{sc} contributions of the elements listed above are shown.

The main contribution on the main path is given by the B1p quadrants and the B1 photodiodes and it results $f_r = 3.22 \cdot 10^{-14}$ for SDB2 during O4 and it was $f_r = 2.32 \cdot 10^{-14}$ in O3. These values are four orders of magnitude smaller than what has been measured in O3 (see Table 6.1), meaning that the light re-coupling with the interferometer may enter from the B5 port.

Actually for the B5 path, the overall contribution, summing up the back-scattering coming from the two quadrants, is $f_r = 1.2 \cdot 10^{-6}$ for O4 and $f_r = 3.2 \cdot 10^{-7}$ for O3. The projections are shown in Fig. 6.15 using the B5 transfer function and are compatible with what has been measured in O3.

6.3.5 Estimation of f_r for SNEB and SWEB

The optical design of the SNEB bench in O4 is shown in Fig. 6.12 and SWEB has a similar optical layout. We replaced several splitters on this bench in order to adapt it for the ITF higher power configuration [100]. The end benches receive the beam in transmission from the 3-km arm cavity of the interferometer. The beam passes towards a telescope and then reaches some sensors, i.e. photodiodes and quadrants, used to control the detector. The power ratio α_p is computed with respect to the entrance of the bench.

The most critical elements are the two lenses (L1a and L1b) at the entrance, the lens L2, the sensors and a Silicon beam dump. These elements are the same on SNEB and SWEB benches and the only difference are the anti-reflecting parameters for lenses L1a and L1b as they were produced at different times and not in the same

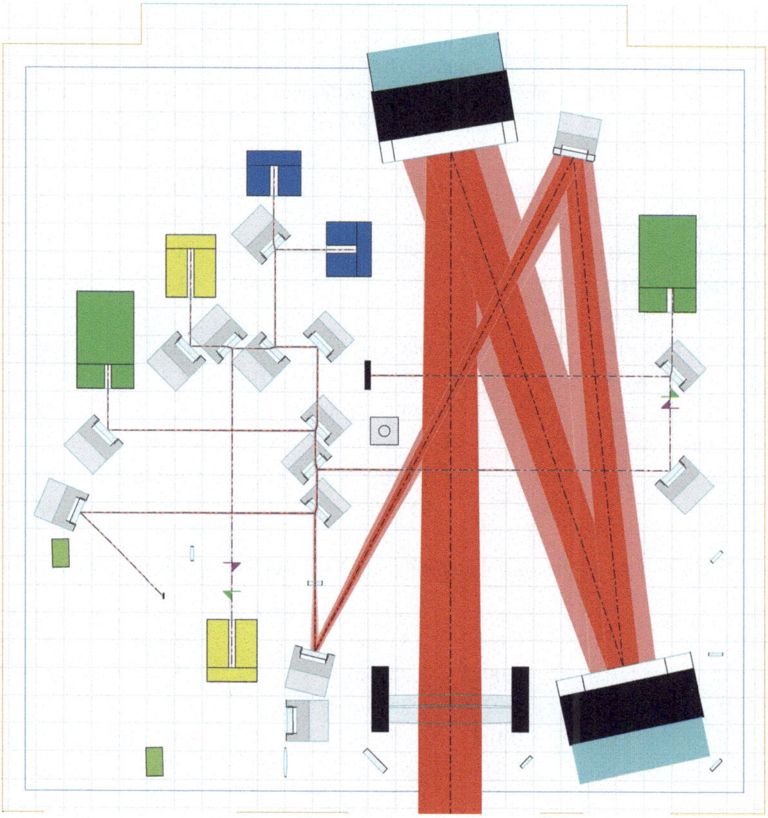

Fig. 6.12 Optocad scheme of SNEB bench. SWEB is similar. The blue boxes represent the photodiodes, the green boxes represent the quadrants and the yellow boxes represent the cameras. The black box near the center of the bench is the Silicon beam dump catching most of the power going to the quadrant. The L1a and L1b lenses are the first elements passed by the beam, at the entrance of the bench

coating run. Table 6.6 shows the parameters and the resulting f_{sp} contribution of the lenses on these benches. The f_{sc} contributions coming from the other elements on the benches are summarized in Table 6.7, where the BRDFs measured at LAPP and reported in Table 6.2 were used.

As was described in Sect. 6.1, the f_r value measured during O3 for SNEB was one order of magnitude bigger than the one of SWEB due to a quadrant shutter open on SNEB. The major contributors to f_r are the lenses L1a and L1b at the entrance of the benches and the quadrants which gives a factor $f_r = 1.7 \cdot 10^{-8}$ for SNEB and $f_r = 2.2 \cdot 10^{-8}$ for SWEB for O4 (assuming the quadrants to be open). In O3 the limiting factor were the lenses L1a and L1b as the quandrants were supposed to be closed (since they were not used). However, during the measurement in O3 the shutter of a quadrant of SNEB was open giving a computed factor $f_r = 5.2 \cdot 10^{-8}$ on SNEB (lenses and one quadrant). For SWEB the shutters were closed so the

Table 6.6 Values of f_{sp} computed for the critical lenses on the bench, i.e. the lenses that might not be tilted, from Eq. 6.12, considering an incidence angle $\beta = 0$. The lenses L1a and L1b at the entrance of SNEB and SWEB are studied. The difference between the two benches is only the reflectivity α since the lenses have been coated at different times. Moreover the lens L2, whose parameters are the same for SNEB and SWEB, is studied. This lens has high f_{sp} if not tilted, so it must be tilted for the run O4

Element	α (ppm)	RoC (m)	z_R (m)	D (m)	f_{sp}
L1a 1st surface (SNEB)	32	2.188	137	1153	$1.6 \cdot 10^{-12}$
L1a 2nd surface (SNEB)	26	−7.334	$2.9 \cdot 10^{-3}$	−6.5	$4.0 \cdot 10^{-10}$
L1b 1st surface (SNEB)	230	−2.979	$1.1 \cdot 10^{-3}$	−3.3	$2.3 \cdot 10^{-9}$
L1b 2nd surface (SNEB)	240	−4.5	$9 \cdot 10^{-3}$	−11.5	$6.1 \cdot 10^{-11}$
L1a 1st surface (SWEB)	200	2.188	137	1153	$9.9 \cdot 10^{-12}$
L1a 2nd surface (SWEB)	270	−7.334	$2.9 \cdot 10^{-3}$	−6.5	$4.2 \cdot 10^{-9}$
L1b 1st surface (SWEB)	320	−2.979	$1.1 \cdot 10^{-3}$	−3.3	$3.0 \cdot 10^{-9}$
L1b 2nd surface (SWEB)	277	−4.5	$9 \cdot 10^{-3}$	−11.5	$7.3 \cdot 10^{-11}$
L2 1st surface	400	$-91 \cdot 10^{-3}$	$1.7 \cdot 10^{-3}$	$-92 \cdot 10^{-3}$	$1.9 \cdot 10^{-4}$
L2 2nd surface	400	$91 \cdot 10^{-3}$	$8.6 \cdot 10^{-3}$	$-243 \cdot 10^{-3}$	$3.7 \cdot 10^{-8}$

Table 6.7 Values of f_{sc} of the optical elements on SNEB and SWEB computed from Eq. 6.7

Element	α_p O3	α_p O4	$w(z)$	$BRDF(\theta)$	f_{sc} O3	f_{sc} O4
B7/B8 QD1	0.08	0.0032	0.35 mm	3	$5 \cdot 10^{-8}$	$8.8 \cdot 10^{-9}$
B7/B8 QD2	0.0736	0.0025	0.35 mm	3	$5 \cdot 10^{-8}$	$5.6 \cdot 10^{-9}$
B7/B8 PD1	0.788	0.02	0.35 mm	$4.9 \cdot 10^{-5}$	$9 \cdot 10^{-11}$	$5.8 \cdot 10^{-14}$
B7/B8 PD2	0.0575	0.108	0.35 mm	$4.9 \cdot 10^{-5}$	$4.8 \cdot 10^{-13}$	$1.7 \cdot 10^{-12}$
B7/B8 beam dump	None	0.815	0.8 mm	$1 \cdot 10^{-4}$	None	$3.7 \cdot 10^{-11}$

computed value for O3 is $f_r = 7.2 \cdot 10^{-9}$. The comparison between measurements and computations for O3 are therefore in agreement (see Table 6.1). For O4, if the quadrants are needed, they will be the limiting element on these benches.

6.3.6 Estimation of f_r for SPRB

The optical design of the SPRB bench is shown in Fig. 6.13 for O3 and O4. It receives the pick-off of the power recycling cavity. The power ratio α_p is computed with respect to the entrance of the bench. The most critical elements are the two lenses (L1a and L1b) at the entrance, the quadrants and the photodiodes.

The f_{sc} contributions coming from the elements on this bench are shown in Table 6.8 and the f_{sp} contributions are reported in Table 6.9.

The f_r value was not measured during O3, but according to the results shown in Tables 6.8 and 6.9, this factor should be $f_r = 1.63 \cdot 10^{-7}$. Despite this value is higher than on the other benches, the transfer function at the SPRB port is less critical than the others, keeping low the projection in Eq. 6.5.

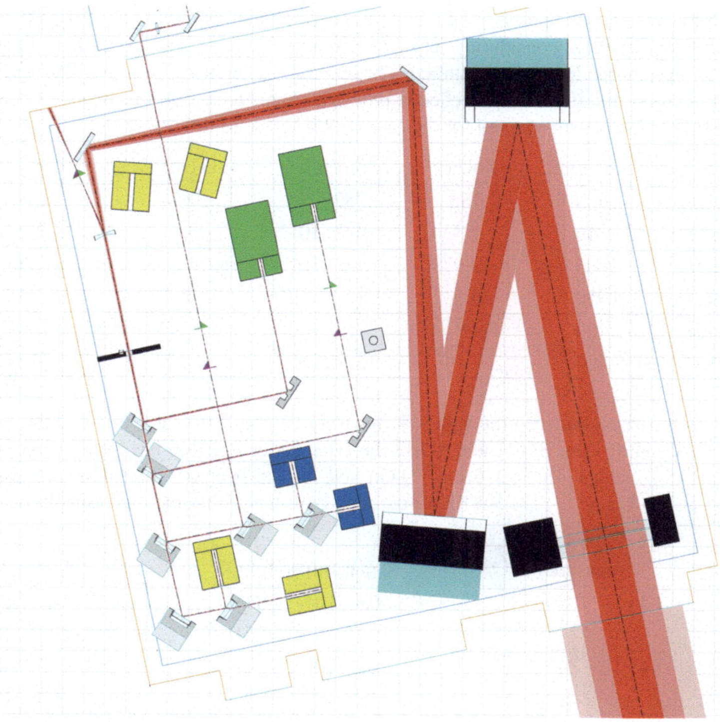

Fig. 6.13 Optocad scheme of SPRB bench. The blue boxes represent the photodiodes, the green boxes represent the quadrants and the yellow boxes represent the cameras. The L1a and L1b lenses are the first elements passed by the beam, at the entrance of the bench

Table 6.8 Values of f_{sc} of the optical elements on SPRB computed from Eq. 6.7

Element	α_p	$w(z)$	$BRDF(\theta)$	f_{sc}
B4 QD1	0.08	0.35 mm	3	$5.7 \cdot 10^{-8}$
B4 QD2	0.08	0.35 mm	3	$5.7 \cdot 10^{-8}$
B4 PD1	0.35	0.35 mm	$4.9 \cdot 10^{-5}$	$1.8 \cdot 10^{-11}$
B4 PD2	0.35	0.35 mm	$4.9 \cdot 10^{-5}$	$1.8 \cdot 10^{-11}$

6.3.7 Noise Projections and Conclusions

The results obtained from the estimates of the f_r parameters for each bench (for the optical configurations of both O3 and O4) are shown in Table 6.10 where, if possible, they are compared with the measurements performed during O3.

As a first result, it has been confirmed that on SDB1, SNEB/SWEB and SPRB, one of the limiting factors are the specular f_{sp} reflections of the lenses. These factors are strongly angle-dependent, and in Fig. 6.14 the trend of f_{sp} as a function of the tilt angle of the lenses is shown. The continuous lines refer to the first surface and the dashed line to the second surface.

Table 6.9 Values of f_{sp} computed for the critical lenses on the bench, i.e. the lenses that might not be tilted, from Eq. 6.12, considering an incidence angle $\beta = 0$. The lenses L1a, L1b at the entrance of SPRB and the lens L2 are studied. The lens L2 has an high f_{sp} if not tilted and it was tilted during O3 by 6° but it has been needed to reduce its angle to limit B4 beam astigmatism. (*) in this case, Eq. 6.12 has been adapted for the case of RoC close to infinity

Element	α (ppm)	RoC (m)	z_R (m)	D (m)	f_{sp}
L1a 1st surface	680	2.188	282	1384	$6.5 \cdot 10^{-11}$
L1a 2nd surface	740	−7.334	$4.0 \cdot 10^{-3}$	−6.5	$2.2 \cdot 10^{-8}$
L1b 1st surface	1600	−2.979	$1.6 \cdot 10^{-3}$	−3.3	$2.9 \cdot 10^{-8}$
L1b 2nd surface	1050	−4.5	$12.5 \cdot 10^{-3}$	−11.5	$5.4 \cdot 10^{-10}$
L2 1st surface	400	flat*	$2.4 \cdot 10^{-3}$	$-148 \cdot 10^{-3}$	$2.6 \cdot 10^{-4}$
L2 2nd surface	400	$73 \cdot 10^{-3}$	$3.5 \cdot 10^{-3}$	$-209 \cdot 10^{-3}$	$7.5 \cdot 10^{-9}$

Table 6.10 Comparison between the values f_r measured during O3 (Table 6.1) and the estimate done for O3 and O4. (*) For SDB1 and SDB2 the B1 transfer function has been used to extract the measured O3 f_r values (see Sect. 6.1). (**) On SNEB it has been considered that one quadrant shutter was open in O3. (***) On SWEB it has been considered that the quadrant shutters were both closed

Bench	f_r measured in O3	f_r estimate for O3	f_r estimate for O4
SDB1	$2 \cdot 10^{-9}$ (*)	$6.4 \cdot 10^{-10}$ (B1)	$1.5 \cdot 10^{-9}$ (B1)
		$4.1 \cdot 10^{-10}$ (B5)	$4.1 \cdot 10^{-10}$ (B5)
SDB2	$4 \cdot 10^{-10}$ (*)	$2.3 \cdot 10^{-14}$ (B1)	$3.2 \cdot 10^{-14}$ (B1)
		$3.2 \cdot 10^{-7}$ (B5)	$1.2 \cdot 10^{-6}$ (B5)
SNEB	$3 \cdot 10^{-8}$	$5.2 \cdot 10^{-8}$ (**)	$1.67 \cdot 10^{-8}$
SWEB	$5 \cdot 10^{-9}$	$7.7 \cdot 10^{-9}$ (***)	$2.16 \cdot 10^{-8}$
SPRB	–	$1.6 \cdot 10^{-7}$	$1.6 \cdot 10^{-7}$

The large input lenses of SDB1 (L1) and SWEB/SNEB (L1a and L1b) are perpendicular to the beam within the mechanical precision. Their contribution is of the same order of magnitude of the most critical elements if they are not tilted. One solution could be to mount them on motorized mechanical supports that allow to tilt them remotely by minimizing their impact on the sensitivity curve. The L2 lens on SWEB (and SNEB) was not tilted during O3, except for an error in the positioning. The first surface of this lens gives an $f_{sp} \sim 10^{-4}$ but it drops very fast with the angle. Since we don't see its contribution on the measurements performed in O3, we assume that this lens is tilted by at least a fraction of deg. The second surface of this lens (dashed light blue in Fig. 6.14) may contribute to the f_r of SWEB (or SNEB). Therefore, this lens should be tilted by $\beta > 0.5$ deg to be sure that this element will not introduce scattered light. On SPRB, the lens L2 was tilted by 6° and therefore it was not contributing to the scattered light noise. However, during the commissioning of O4 it has been required to reduce β of L2 to reduce the astigmatism of the beam B4. In order to be safe (keeping $f_r < 10^{-7}$), the tilt angle should be $\beta > 2$ deg.

The projections of scattered light noise, calculated from Eq. 6.5, using the f_r factors measured in O3 and those estimated for O4, are shown in Fig. 6.15.

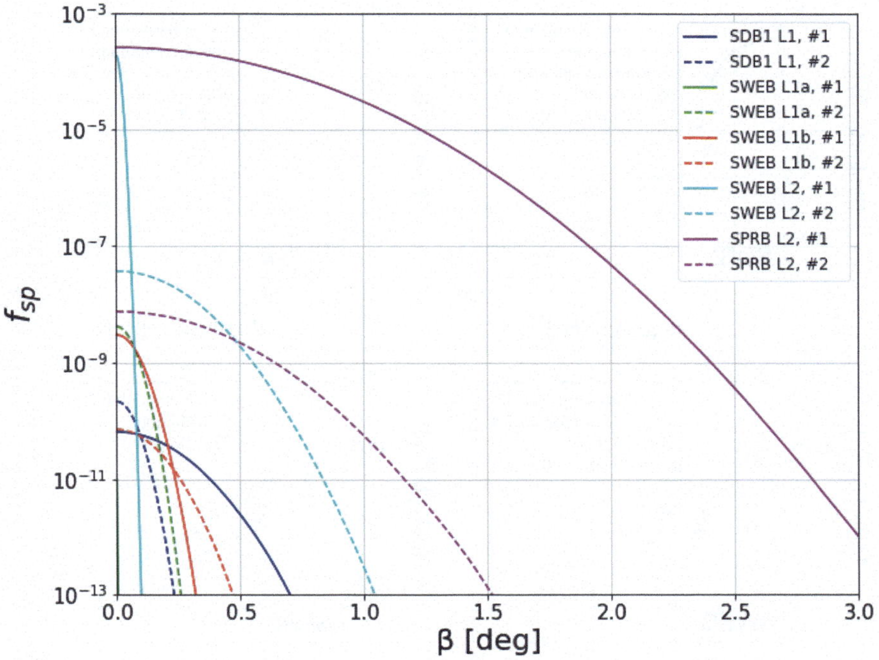

Fig. 6.14 Specular reflection of the most critical lenses of the benches as a function of the tilt angle: the meniscus lens on SDB1 (L1), the lenses L1a and L1b on SWEB (similar for SNEB), the lens L2 on SWEB (similar for SNEB) and the lens L2 on SPRB. The continuous lines refer to the lenses first surface and the dashed line to the second surface

From the comparison between the projections using the values of f_r measured in O3 and calculated for O4, it can be concluded that the effect of the scattered light coming from SDB1 and SDB2 will be the same as that measured during O3. The scattering noise can worsen therefore the sensitivity of the detector up to ~15 Hz. The limiting elements on SDB1 main path are the TGG crystal, the OMC cavity and the meniscus lens. Concerning the meniscus lens, it is planned to install it on a remotely rotating mount in the future, so as to reduce the coupling of its back-scattering with the interferometer beam. However, it is complicated to improve the situation as it is not trivial to change the TGG crystal and OMC cavity with elements that scatter less. What will be done is to measure at LAPP (see Sect. 6.2.1) their BRDFs, since in these calculations we used values found in literature. One option to reduce the effect of the scattered light from SDB1 would then be to act on the relative movement between the scatterer and the interferometer, $z_r(t)$. In the fringe wrapping approximation the observable to act on is the speed of the bench to reduce the cut-off frequency f_{max} below 10 Hz (see approximation in Eq. 6.14). To do this, one must act at the suspension level.

The scattered light noise from SDB2 is dominated by the back-scattering of the B5 beam. This value is expected to be a factor of 2 higher in O4 than in O3, due

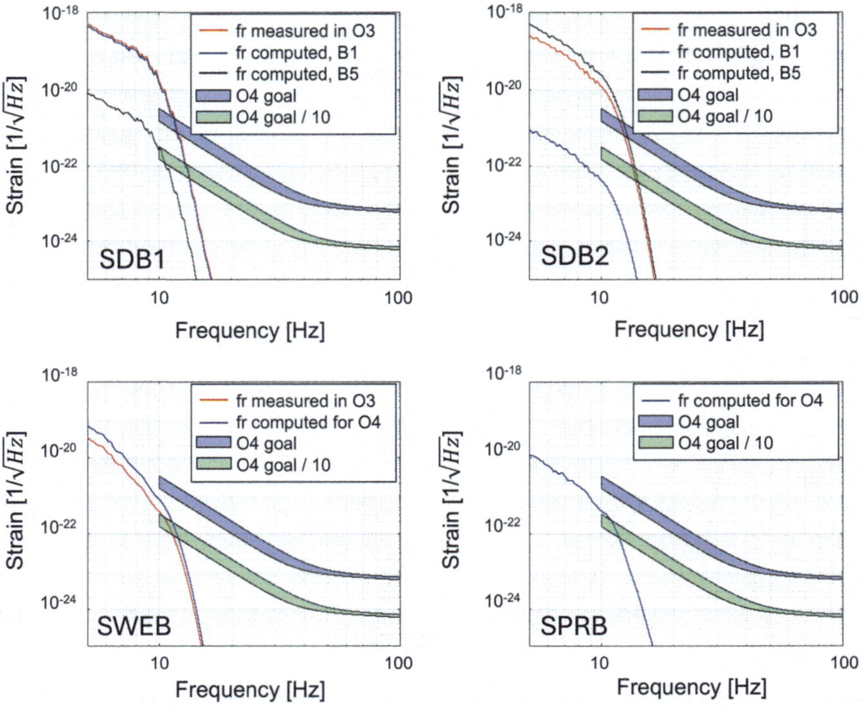

Fig. 6.15 Scattered light noise projections for O4 for different suspended benches (SDB1 up left, SDB2 up right, SWEB down left and SPRB down right). Different values for the re-coupling factor f_r have been used and compared: f_r measured during O3 where available (curves in red) and f_r calculated for O4 (curves in blue for the main beam referred as B1 beam and curves in black for the B5 beam). For all these projections the O4 transfer functions have been used

to the replacement of the beam splitters on the bench. Since the sensitivity can be spoiled up to 15 Hz, one solution will be to replace the B5 quadrants on SDB2 with others whose BRDF is smaller.

Concerning SWEB, the projection for O4 is a factor of 2 above the one that uses the factor f_r of O3 (considering that the quadrants will not be shuttered during O4). The limiting elements on SNEB, SWEB and SPRB are the lenses L1a and L1b and the quadrants. The solution to reduce the factor f_r arising from these benches is therefore to place the lenses on a motorized mount, as said for L1 on SDB1, and to replace the quadrants with less scattering ones. The effect of the terminal benches and SPRB, would worsen only the optimistic estimate of the detector sensitivity, which is 10 times better than the O4 goal. This means that the intervention on these benches is less urgent than on SDB1 and SDB2.

6.4 Conclusion

This chapter describes scattered light noise and how it worsens the sensitivity of gravitational wave detectors. Scattered light noise coming from Virgo suspended benches is one of the major sources of this noise. The scattered light is generated by the optics on the benches and is modulated by their relative motion with respect to the interferometer. This chapter reports the measurements of the scattered light noise coming from the suspended benches performed during O3 where it was demonstrated that this noise worsen the low frequencies. This would prevent to notice the low frequency sensitivity improvement that will be introduced by the frequency-dependent squeezing. Hence, we studied the scattered light noise projections for O4 to check its impact on Virgo sensitivity. To obtain accurate estimates of the amount of light scattered by the optical elements on the benches, we built an interferometric scattermeter at LAPP to perform such measurements. The result of the study shows that the suspended benches of the detection system can limit the sensitivity of Virgo during O4. The limiting elements on the benches have been identified and this gives the possibility to mitigate the most critical sources. The next step is to measure the scattered light noise arising from each bench by using the sensitivity curve of Virgo and compare it with the simulations in order to identify possible additional sources of noise.

Chapter 7
Ghost Beams Study and Mitigation

The interferometer's optical benches have a complex optical layout and contain many elements. These elements can create spurious beams, e.g. due to non-perfect anti-reflecting coatings, which propagate on the benches. These beams are called *ghost beams*. It is possible to trace them using a software called *Optocad* [84], which is a Fortran 95 module for tracing Gaussian TEM00 (defined in Appendix A) beams propagation within an optical set-up. In Fig. 7.1 the propagation of a ghost beam on Optocad is shown. The incoming beam propagates and it is almost completely reflected by a mirror. However, a fraction of the beam is transmitted from the outer surface as the Highly Reflective (HR) coating has residual transmission (of the order of a few ppm). Subsequently, this beam arrives at the back surface and is partially reflected by Anti-Reflective (AR) coatings (a fraction of hundreds of ppm). The ghost beam generated in this way leaves the first surface and propagates on the optical bench where it can be scattered by diffusing elements, such as mirror mounts. The scattered light thus generated can re-couple with the main beam of the interferometer, introducing the noise described in Eq. 6.5 through the factor f_{extra} in Eq. 6.6.

The approach followed was to trace the ghost beams from each optical element of the suspended benches with Optocad and calculate their power from the residual transmissivity η_T and reflectivity η_R of the optics [101–103]. For example the power of the ghost beam in Fig. 7.1 can be computed from

$$P_{GB} = \eta_R \cdot \eta_T^2 \cdot P_{in} \tag{7.1}$$

where P_{in} is the input beam power. An estimate of the fraction of the power of a ghost beam that re-couples with the main interferometer beam is given by

$$f_{extra} = \frac{P_{GB}}{P_{in}} \cdot f_{sc}(Al) \tag{7.2}$$

E. Polini, *Broadband Quantum Noise Reduction in Advanced Virgo Plus*,
Springer Theses, https://doi.org/10.1007/978-3-031-95143-5_7

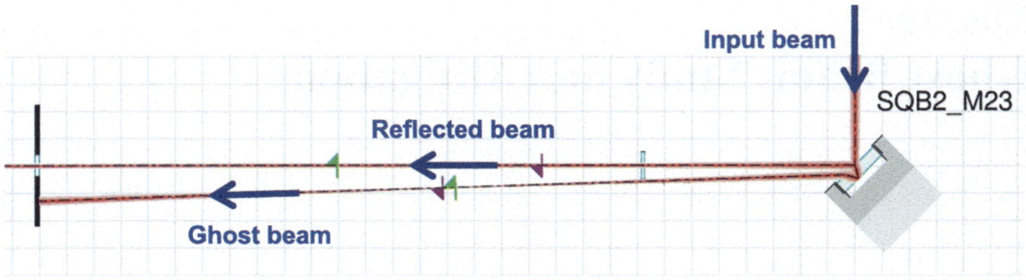

Fig. 7.1 A beam propagating towards a mirror. A ghost beam is generated at the mirror surfaces. A diaphragm lets the main reflected beam propagate and it dumps the ghost beam

where $f_{sc}(Al)$ is the scattered light from the aluminum given by Eq. 6.7, which well approximates the scattering elements on the benches.

7.1 Mitigation Strategies

It has been decided to dump the beams that could hit critical elements, such as highly diffusing ones, or the more powerful ones by using absorbing surfaces along their path. Several mitigation strategies have been adopted, depending on several constraints such as space and weight availability on suspended benches. The mitigation strategies used on all the benches of the Virgo detector during commissioning for run O4 are described below.

7.1.1 Wedges Optimisation

Most bench optics have a wedge, i.e. the two surfaces of the component are not parallel but have a small angle between them, called wedge angle. This is so that the ghost beam coming out of the optical component is angularly separated from the reflected beam, so that it can be better dumped. The wedge can be oriented in any direction, but we have chosen to orient it horizontally so that the ghost beams propagate parallel to the bench plane and are easier to stop. Figure 7.2 shows for example the ghost beam generated by the mirror SQB1_M21 for different wedge orientations (on the left $-0.5°$ and on the right $+0.5°$) and in the absence of the wedge (in the middle). The optimal wedge angle in this case is $-0.5°$ as the ghost beam is stopped by an already planned diaphragm, whereas with a wedge of $0.5°$ it would have been very difficult to stop it due to space constrains. In the absence of a wedge, the ghost beam would instead propagate inside the Faraday Isolators and risk to produce additional scattered light. Using simulations on Optocad, we optimized the wedge of all the optics on Virgo's benches so that the most powerful ghost beams could be stopped.

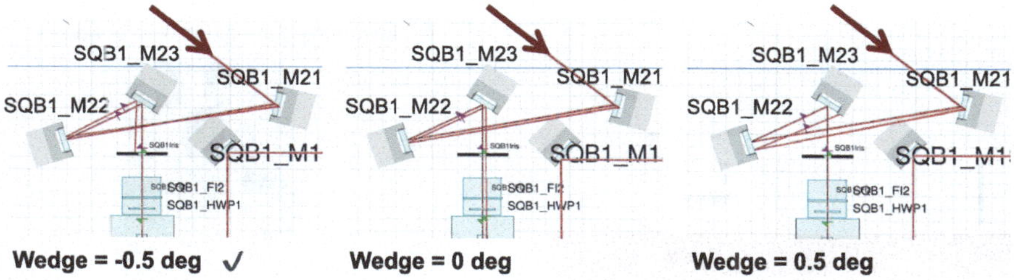

Fig. 7.2 Example of ghost beam generated at the SQB1_M21 surfaces for different values of the wedge: $-0.5°$ (left), $0°$ (middle) and $0.5°$ (right)

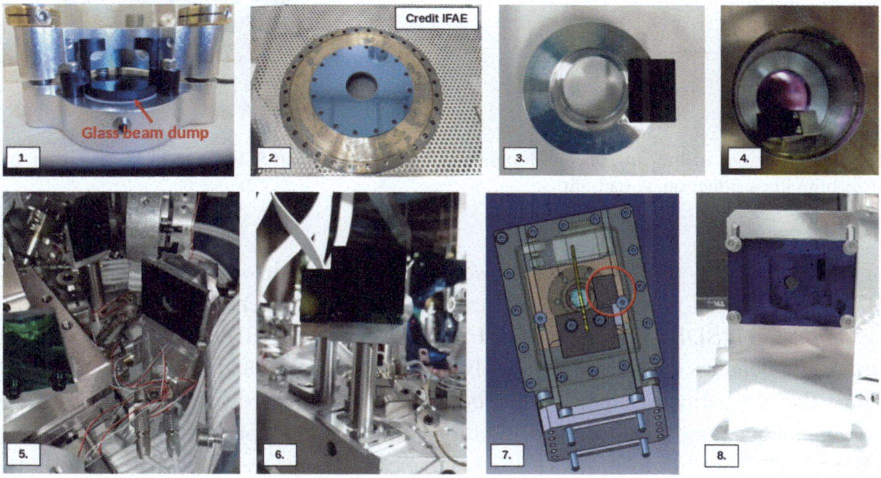

Fig. 7.3 Some of the dumping strategies implemented in AdV+ during the commissioning: (1) absorbing glass behind mirrors, (2) baffles inside tubes, (3) absorbing glass glued on a lens mount, (4) beam dump in front of a viewport, (5) absorbing glass glued on a quadrant and few other beam dumps, (6) small beam dump with black absorbing screws, (7) beam dump glued on the photodiode box, (8) diaphragm

7.1.2 Dumping Elements

After tracking the ghost beams and identifying those with the highest power or going in the most dangerous directions, we decided how to stop them. First of all, we had to optimize the number of beam dumps due to space and weight constraints of the suspended benches, i.e. we made sure that one beam dump could stop several spurious beams. Several beam dumps are shown in Fig. 7.3: the small beam dumps with black absorbing screws (6–4), diaphragms (8) and baffles (2) inserted inside the tubes (more on the design of the latter two are given later). In addition, when there was no space we glued absorbing glass onto the mounts of the elements, as shown in Fig. 7.3 (on the mount of a lens (3), on a quadrant (5), on the in air box of a photodiode (7)). To stop the transmitting spurious beam from the second surface

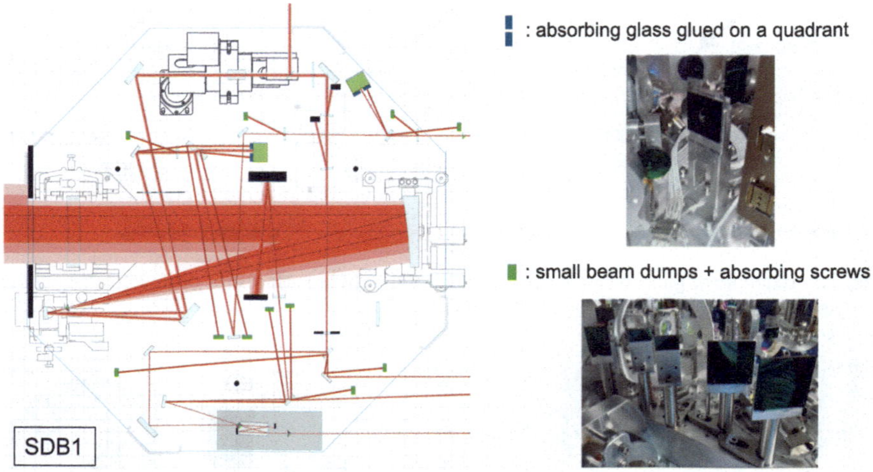

Fig. 7.4 Positions of the small beam dumps installed on the bench SDB1 and the example of an absorbing glass glued on a quadrant

of the mirrors, absorbing discs were inserted behind the mirrors, as shown in Fig. 7.3 (1). The study of ghost beams was carried out on the benches of the detection system (SDB1, SDB2, SNEB and SWEB) [102], injection (SIB1, SIB2, EIB) [103], squeezing benches (EQB1, SQB1, SQB2), in the link tubes and filter cavities [101]. Thus, the mitigation strategies described were implemented in all these cases. In Fig. 7.4, the example of some mitigations implemented on the detection bench SDB1 is shown: the positions of the small beam dumps mounted with absorbing screws and of the absorbing glasses glued on the quadrants. The diaphragm and baffle optimization will be detailed in Sect. 7.2.

7.2 Diaphragm and Baffle Design

The diaphragms shown in Fig. 7.3 (8) were designed and installed on the SQB1 and SQB2 benches and the baffles shown in Fig. 7.3 (2) inside the pipes and filter cavity of the FDS system (see Fig. 4.5). Actually, these benches have less stringent space constraints and by using diaphragms, it is possible to stop ghost beams very efficiently. However, SQB1 and SQB2 present an additional difficulty in that the beams propagate in two directions: towards the filter cavity and towards the interferometer, as shown in Figs. 4.11 and 4.12. This means that each surface generates two ghost beams. For this reason the absorbing glasses are installed on both the sides of the diaphragm, to catch ghost beams in both directions and scattered light.

In order not to introduce propagation losses in the FDS system, it was necessary to study the optimal size of aperture of the diaphragms and baffles. A Gaussian beam propagating through a circular aperture suffers from power losses, that can be seen as optical losses, and diffraction losses, which represent mode mismatch losses. Since the squeezed beam is very sensitive to these losses, it is important to keep them under a safety threshold. Since the squeezed beam propagating over the system is subject to few ppm of optical losses due to residual transmission of the AR coatings of the mirrors, the threshold for losses due to the clipping of the Gaussian beam tails across an aperture has been set to 1 ppm. Moreover, the requirement for the mode matching (see Appendix A) losses on the FDS system is about 1%. Considering a factor 10 of margin, the mode mismatch introduced by the diffraction of the Gaussian beam propagating through all the circular apertures must be less than 0.1%. If we consider an order of 10 dumping elements on the beam path, the mode mismatch introduced by each aperture mush stay below 0.01%. We designed the diaphragms and the baffles taking into account those specifications as detailed in Sects. 7.2.1 and 7.2.2.

More details on this study can be found in [104].

7.2.1 Power Losses Through a Circular Aperture

The transmitted power of a Gaussian beam propagating along z through a circular aperture of radius a is given by

$$P_T(a, z) = P_0 \left[1 - \exp\left(-\frac{2a^2}{w(z)^2}\right)\right] \tag{7.3}$$

where P_0 is the incident power and $w(z)$ is the beam radius at the aperture plane.

As we cannot be sure that the beam is perfectly centered, it is necessary to study what happens if the Gaussian beam is laterally misaligned by a quantity d with respect to the axis of the diaphragm's or baffle's aperture of radius a, as shown in Fig. 7.5.

The transmitted power of a Gaussian beam, propagating along z through a circular aperture of radius a and laterally displaced by the quantity d is given by [105]

$$P_T(a, z) = P_0 \exp\left(-\frac{2d^2}{w(z)^2}\right) \sum_{k=0}^{\infty} \frac{2^k d^{2k}}{w(z)^{2k}(k!)^2} \left[\gamma\left(k+1, \frac{2a^2}{w(z)^2}\right)\right] \tag{7.4}$$

where γ is the lower incomplete Gamma Function

$$\gamma\left(k+1, \frac{2a^2}{w(z)^2}\right) = \Gamma(k+1) - \Gamma\left(k+1, \frac{2a^2}{w(z)^2}\right) \tag{7.5}$$

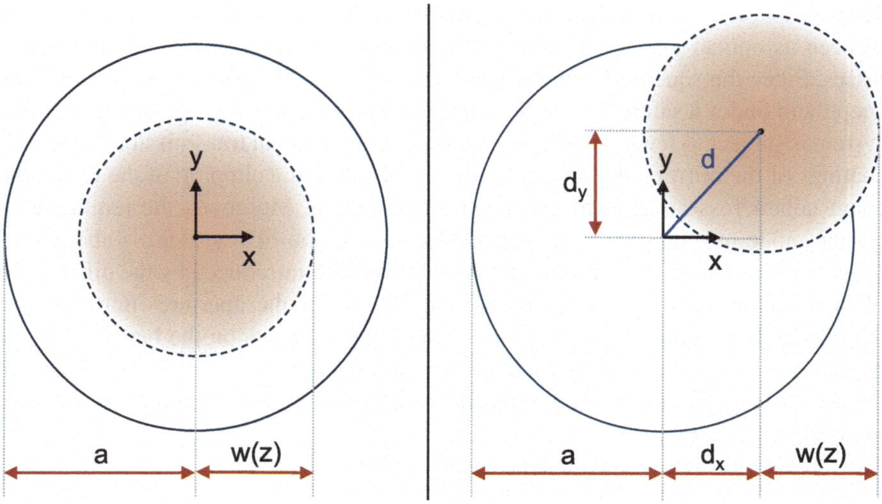

Fig. 7.5 Gaussian beam propagating through a circular aperture of radius a: on the left the beam direction is well aligned with the aperture axis, on the right the beam is displaced by a quantity d

where $\Gamma(z)$ is the gamma function and $\Gamma(z, x)$ incomplete gamma function. The optical losses of the Gaussian beam propagating through the aperture normalized for the incident beam power as a function of the ratio between the aperture a and beam $w(z)$ radii, for different values of the lateral displacement d are shown in Fig. 7.6. In the absence of beam displacement, i.e. $d \sim 0 \cdot w(z)$, the power loss requirements are satisfied for a ratio $\frac{a}{w(z)} > 2.6$. However, in the FDS system, we expect displacements of $d \sim 1 \cdot w(z)$ for the diaphragm and $d \sim 1.5 \cdot w(z)$ for the baffles. Figure 7.6 shows that to be within the requirement we should consider $\frac{a}{w(z)} > 3.5$ for the diaphragm aperture and $\frac{a}{w(z)} > 4$ for the baffles.

7.2.2 Diffraction Losses Through a Circular Aperture

A Gaussian beam can be weakly diffracted by a circular aperture.

The diffracted beam from the aperture can be approximated to a Gaussian beam because we are in the far-field approximation [26]. Considering therefore that the beam incident on the aperture has a waist w_0, we denote by w'_0 the waist of the diffracted beam. The ratio between these two parameters is given by [106]

$$1 - \frac{w'_0}{w_0} \sim \exp\left(-\frac{2a^2}{w(z)^2}\right) \cos\left(p \frac{a^2}{2\pi\, w(z)^2}\right) \qquad \text{with} \qquad p = \frac{\sqrt{2}z\lambda}{w_0^2} \tag{7.6}$$

where $p = 0$ corresponds to the case where the waist is on the aperture plane. Figure 7.7 shows different curves for different values of the parameter $p =$

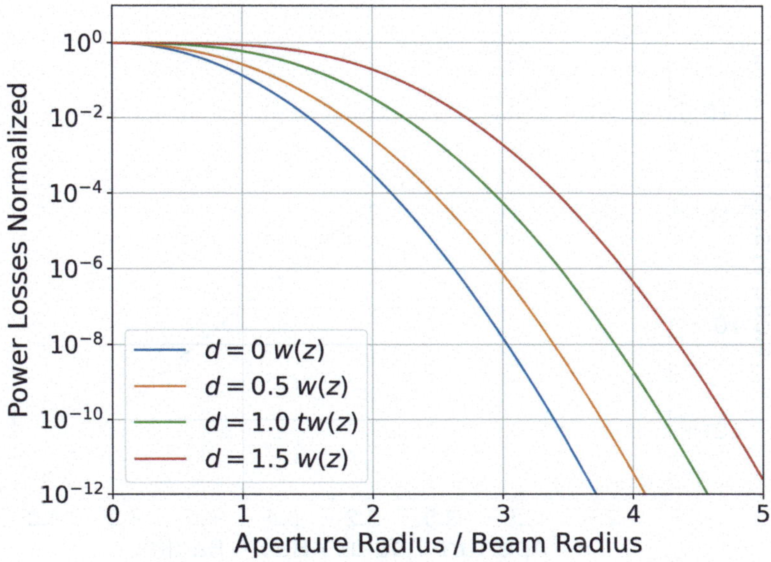

Fig. 7.6 Normalized power losses as a function of a lateral displacement $d = 0.0, 0.5, 1.0, 1.5 \cdot w(z)$

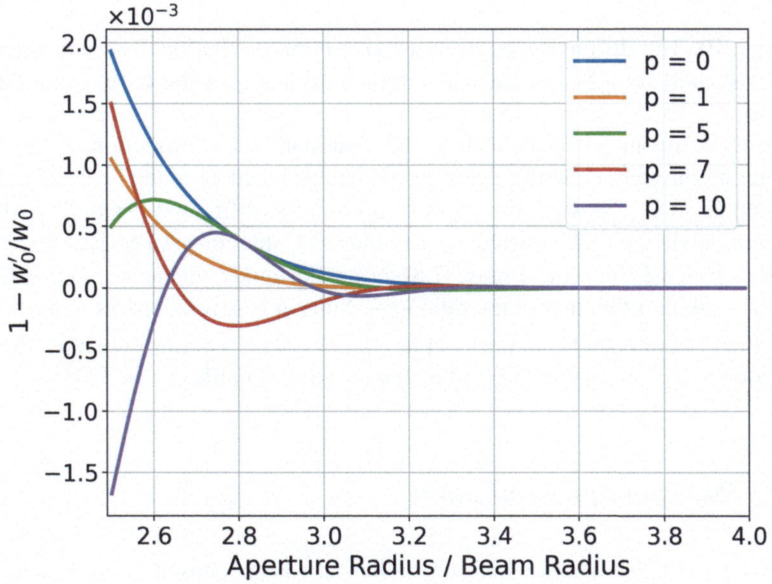

Fig. 7.7 Difference between diffracted and incident beam waists as a function of the ratio between the aperture radius and the beam size, for different values of the parameter p

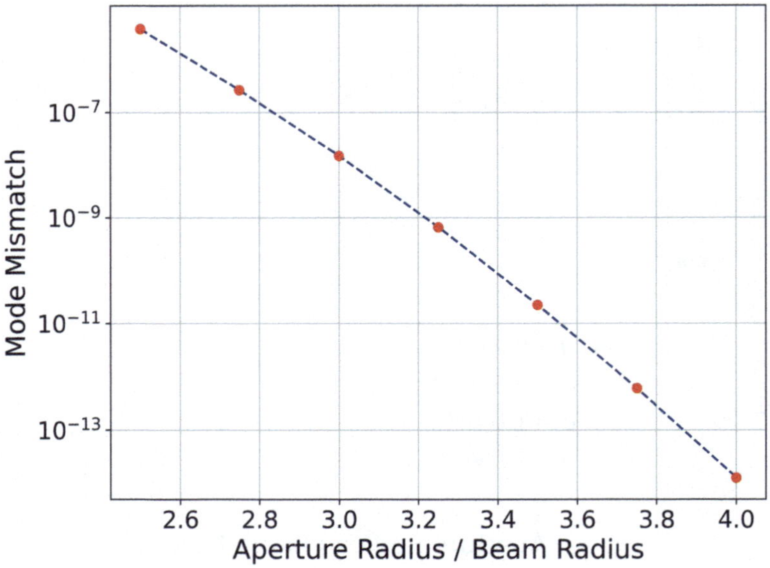

Fig. 7.8 Mode Mismatch as a function of the ratio between the aperture and incident beam radii for $p < 10^{-3}$

0, 1, 5, 7, 10. The difference in the waist size between the incident and diffracted beams, is small for $\frac{a}{w(z)} > 2.8$ where the contribution of the cosine starts to get smaller.

The mode mismatch associated to the Gaussian beam diffraction of the beam propagating through a circular aperture can be estimated from the Overlap Integral (OI) (see Eq. 6.11) between the input beam and the diffracted beam. The OI has been computed with OSCAR (Optical Simulation Containing Ansys Results) [107], a Matlab based FFT code. Figure 7.8 shows the mode mismatch, computed as MM = 1 − OI, as a function of the ratio $\frac{a}{w(z)}$. Since it is several orders of magnitude lower then the threshold at 0.01% for $\frac{a}{w(z)} > 2.6$, we can conclude that the diffraction losses are negligible for the diaphragm and baffles in the FDS system.

7.2.3 *Positioning Optimization*

The position of the diaphragms and baffles has been chosen to minimize their number and to ensure that most of the considered ghost beam is blocked. Since the beams propagate in the two directions, we put the absorbing glass on both sides. In addition to avoiding clipping losses and therefore having chosen an aperture of radius of $a = 3.5 \cdot w(z)$ for the diaphragms and $a = 4 \cdot w(z)$ for the baffles, one

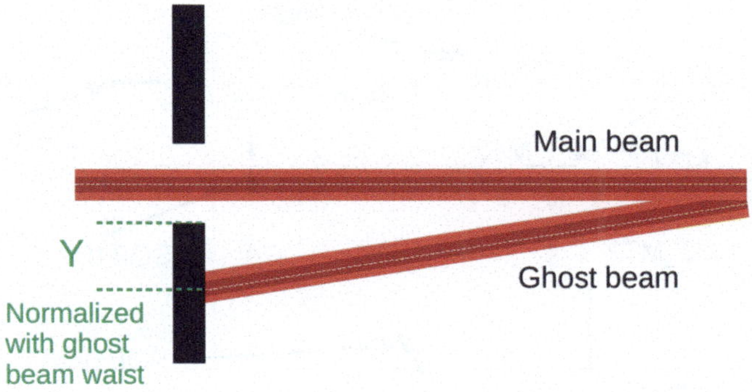

Fig. 7.9 Main beam propagating through an aperture and ghost beam stopped by the diaphragm surface. d represents the distance between the ghost beam and the main beam and d_{edge} represents the distance between the ghost beam and the edge of the aperture hole

must be careful that the ghost beam is contained on the diaphragm. For this purpose we set two constraints. The first one is the distance d between the main beam and the ghost beam (at the diaphragm plane surface). The second one is the distance between the ghost beam and the edge d_{edge} of the aperture hole normalized by the ghost beam size at the aperture $w(z)_{GB}$, defined as $Y = \frac{d_{edge}}{w(z)_{GB}}$ as shown in Fig. 7.9.

The threshold values for the parameters d and Y have been set as

- $d > 7 \cdot w(z)_{average}$ for diaphragm and $d > 8 \cdot w(z)_{average}$ for baffles, where $w(z)_{average}$ is the average between the main and ghost beams size with the same logic used in Sect. 7.2;
- $Y > 3.5$ for diaphragms and $Y > 4$ for baffles, with the same logic used in Sect. 7.2.

The diaphragms were arranged in such a way as to meet these requirements and an example of their arrangement on the SQB1 bench is shown in Fig. 7.10.

Moreover, in order to make the diaphragm manufacturing easier, we have standardized the aperture sizes. Actually, we chose three sizes of apertures in order to meet the requirements of main beam clipping and ghost beam stopping. The radius of the apertures are $a = 4, 5.5, 6.5$ mm.

The diaphragms are absorbing glasses coated with AR coatings, with residual reflectivity measured at LAPP of $\eta_R(diaphragm) < 1\%$ for an incident beam with $\lambda = 1064$ nm and angle of incidence $\leq 10°$. Their external size is 75×50 mm for most of them and 110×50 mm for some others where the distance between the main and ghost beam is bigger, as summarized in Table 7.1.

More details on this analysis can be found in [101].

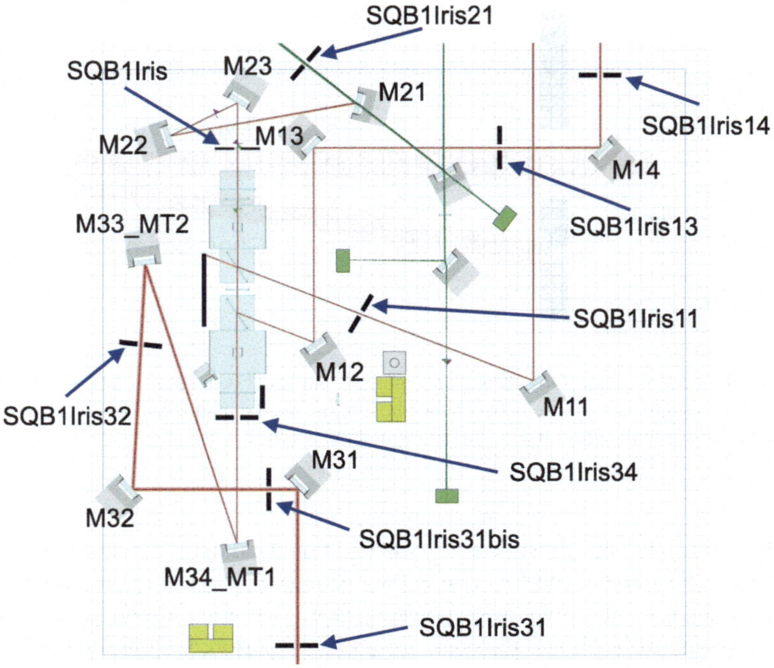

Fig. 7.10 Diaphragms on SQB1 bench

Table 7.1 Dimensions of each diaphragm. We need two for each of them since each of them is coated only on one side but we want to catch light on both faces, to be sure to dump as much stray light as possible. (*) refers to the oblong one shown on the right of Fig. 7.12

Name	Hole diameter (mm)	Height (mm)	Width (mm)
SQB2Iris	8	50	110
SQB2Iris2	11	50	110
SQB1Iris11	8	50	75
SQB1Iris13	8	50	75
SQB1Iris14	8	50	75
SQB1Iris21	Oblong*: 21	50	75
SQB1Iris34	8	50	110
SQB1Iris32	13	50	75
SQB1Iris31bis	11	50	110
SQB1Iris31	11	50	75
SQB1Iris	8	50	75

7.3 Ghost Beams from Filter Cavity Mirrors

The ghost beams arriving from the mirrors of the filter cavity are more powerful than the ones generated by the optics on the benches. This is due both to the residual transmissivities and reflectivities reported in Table 7.2 and to the main beam power amplified inside the cavity.

The incident power (sub-carrier beam) on the external surface of the input mirror (IM) is $P_{in} = 10$ mW. The optical gain of the filter cavity at 1064 nm is given by

$$G_{opt} = \left(\frac{t_{IM}}{1 - r_{IM} \cdot r_{EM}} \right)^2 = \frac{T_{IM}}{(1 - r_{IM} \cdot r_{EM})^2} = 6347, \tag{7.7}$$

from the parameters in Table 7.2. The power circulating inside the cavity (for the subcarrier beam, see Sect. 4.2.3) is

$$P_{circ} = G_{opt} \cdot P_{in} = 63\,\mathrm{W}. \tag{7.8}$$

The four more powerful ghost beams generated by the filter cavity mirrors are shown in Fig. 7.11 and are referred to as

- ghost beams 'a' and 'b' which are going back towards SQB2 and SQB1 benches;
- ghost beams 'c' and 'd' which are propagating inside the cavity.

The power of these ghost beams is computed with the values in Table 7.2 and it is given in Fig. 7.11. The power P_{GBc} has been computed assuming that the

Table 7.2 Reflectance and transmittance for HR and AR coatings for FC input (IM) and end (EM) mirrors

Parameter	@ 1064 nm	@ 532 nm
T_{IM}	620 ppm	6.3%
T_{EM}	5 ppm	6.3%
R_{IM}	200 ppm	200 ppm
R_{EM}	1000 ppm	1000 ppm

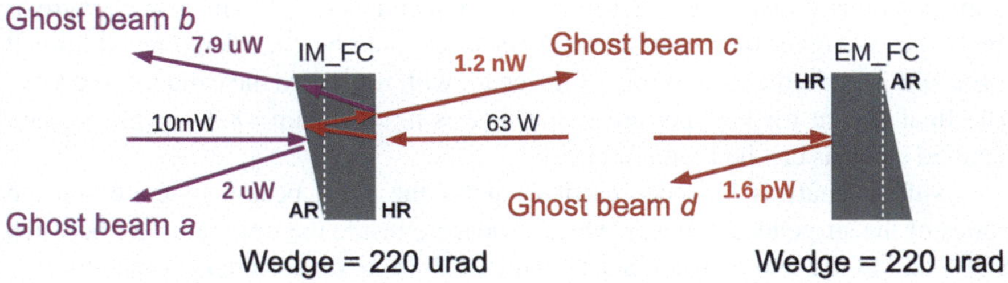

Fig. 7.11 Ghost beams generated by the filter cavity mirrors: the ghost beams 'a' and 'b' are reflected by the cavity and go towards the suspended benches, while the ghost beams 'c' and 'd' propagate inside the filter cavity

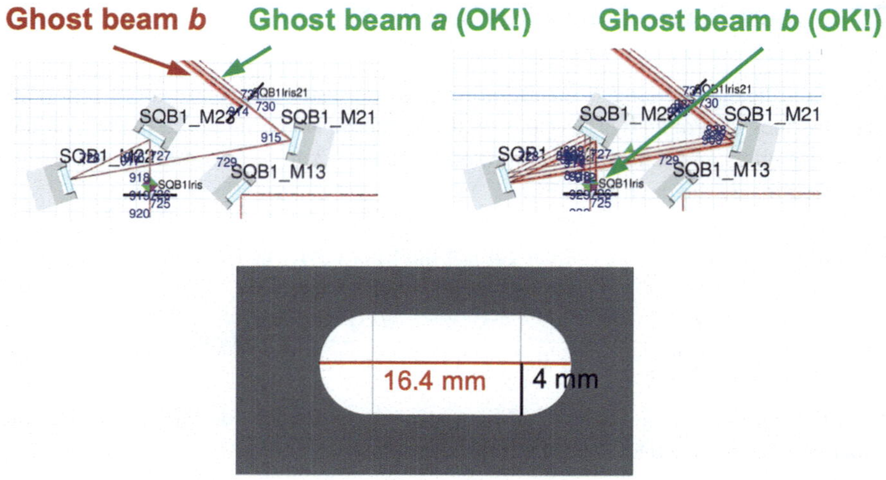

Fig. 7.12 On the top, the ghost beams 'a' and 'b' arriving on SQB1 bench. On the bottom, the asymmetric diaphragm installed on SQB1

beam resonating inside the cavity and the incoming beam interferes constructively inside the input mirror, as an upper limit. This means that the power inside the IM considered is $P_{in} = 10$ mW.

The position of the diaphragms and baffles was chosen so as to contain ghost beams and not clip the incident beam, as described in Sect. 7.2.3 and taking into account space constraints.

The ghost beams 'a' and 'b' are dumped on diaphragms on SQB1 as shown in Fig. 7.12. The diaphragm SQB1Iris21 has been designed asymmetric in order to let the ghost beam 'b' propagate till SQB1Iris in order to be properly dumped.

The ghost beams 'c' and 'd' are stopped by two baffles inside the filter cavity, as shown in Fig. 7.13. The vertical grey lines represent the optimal position for the baffles in order to satisfy the requirements described above: $z_{IM} + 113.8$ m and $z_{EM} - 113.8$ m. For mechanical constrains of the tube, the baffles have been installed at the positions shown on Fig. 7.14 $z_{IM} + 118$ m and $z_{EM} - 118$ m. The aperture of the baffles is bigger than $a = 4 \cdot w(z)$ since it should be considered an additional error of ± 2 cm in the positioning of the beam with respect to the center of the tube. The final choice for the aperture radius size is $a = 75\,\text{mm} = 8.3 \cdot w(z)$. A more detailed analysis can be found in [101].

A similar analysis has been carried out for the ghost beams generated on the optics of the suspended benches which propagate inside the linking tubes: between SQB1 and SQB2 and between SQB2 and FCIM. More details are in Appendix B.

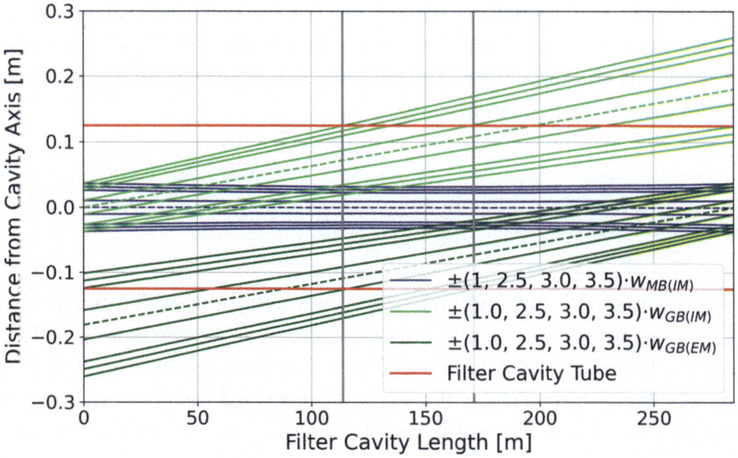

Fig. 7.13 Ghost beams 'c' (light green) and 'd' (dark green) propagating inside the filter cavity next to the main bean (blue). The horizontal red lines represent the filter cavity tube internal surface. The vertical gray lines represent the proposed position for the baffles inside the Filter Cavity, with a positioning precision of ±0.5 m. The power of the ghost beams is 1 nW

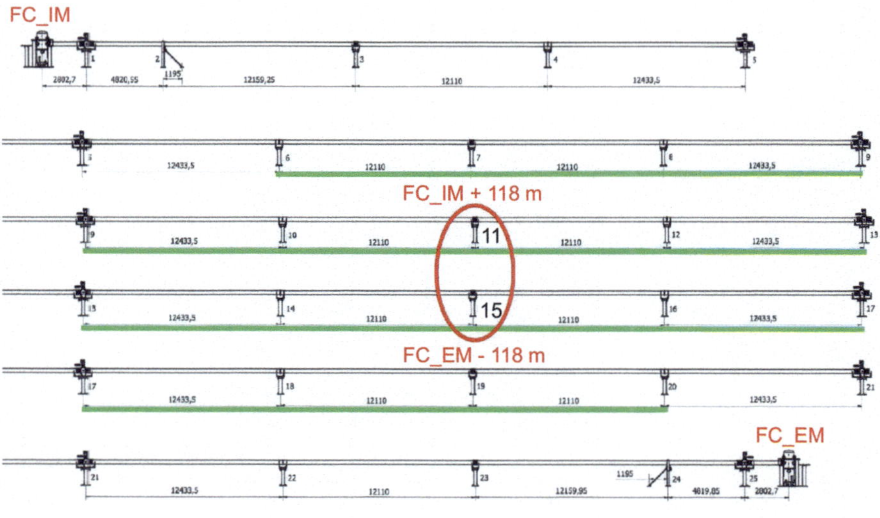

Fig. 7.14 Filter cavity structure with the proposed position for the baffles, placed at $z_{IM} + 118$ m and $z_{EM} - 118$ m, in correspondence of the supports 11 and 15 respectively

7.4 Ghost Beam Study on the Green Path

To complete the study of ghost beams on the FDS system, we studied also the ghost beams on the green path [101]. The motivation is to prevent these beams from reaching the PSDs that will be used to control SQB1 and SQB2, see Sect. 5.2.2, in order to avoid the degradation of the control loops accuracy. The wedges of the mirrors have been optimized and it has been decided to add diaphragms in front of the PSDs with AR coating on the external side. The external size for the diaphragms on the green path is 75×50 mm.

7.5 Conclusion

In this chapter, an other contribution to the scattered light noise has been considered. Spurious beams generated by the imperfect coatings of the optics, called ghost beams, can be scattered and can recouple with the main interferometer beam, introducing noise. We have traced the ghost beams generated on Virgo's optical benches and developed dumping and mitigation strategies, such as wedges optimization, diaphragms and baffle design. We have optimized positions and the aperture size of the diaphragms to minimize losses and efficiently dump the ghost beams.

Chapter 8
Active Control of Scattered Light on the Frequency Dependent Squeezing System

Scattered light has been observed to be an issue to measure the frequency dependent squeezing at the needed frequency on the homodyne detector on EQB1 on the FDS system, as mentioned in Sect. 5.6. In this chapter I will explain how the scattered light evidence has been found and how this problem was mitigated.

8.1 Scattered Light Evidence

The first evidence of scattered light was found by sending only the LO on the EQB1 homodyne detector. By measuring the shot noise on the differential audio channel of the homodyne, we noticed an excess of noise, a *bump* at low frequencies (the orange curve in Fig. 8.1).

We investigated the origin of this noise by studying the spectrogram of the homodyne channel and the interference fringes in the time domain. The spectrogram shows typical stray light arches and the time series of the signal the typical interference structure, as shown on the right side of Fig. 8.1. To confirm that this was scattered light noise, we inserted the delay line on EQB1, which prevents scattered light from propagating to the filter cavity. In this case there is no evidence of the low-frequency bump, as shown by the blue curve on the differential audio channel in Fig. 8.1. In this case the residual motion between the scattering object and the suspended elements is $z_r(t) = \text{constant}$, meaning that there is no scattered light noise. So the conclusion of this qualitative analysis is that there is a scattering element on the system producing a scattering field E_{sc}. As illustrated in Fig. 8.1 this field propagates on the suspended optics, i.e. up to the input mirror of the filter cavity (FCIM) that reflects it back to the homodyne detector. E_{sc} sees an optical path modulation $z_r(t)$ due to the fact that EQB1 is attached to the ground and the FCIM is suspended.

E. Polini, *Broadband Quantum Noise Reduction in Advanced Virgo Plus*,
Springer Theses, https://doi.org/10.1007/978-3-031-95143-5_8

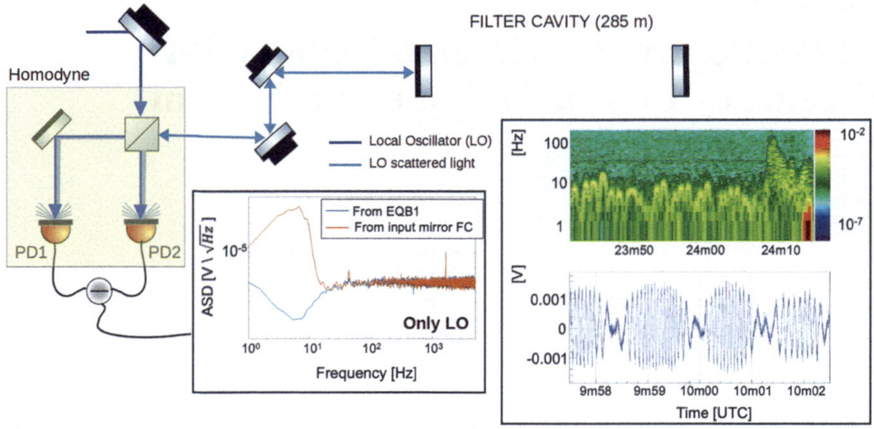

Fig. 8.1 Simplified scheme of the effect of the scattered light from the homodyne photodiodes (in light blue) re-coupling with the local oscillator (in blue) producing a low-frequency bump on the audio differential signal (orange line on the graph on the left). On the top right, the spectrogram of the homodyne differential signal shows the arc structures typical of scattered light while on the bottom right the homodyne signal in the time domain shows the LO's interference fringes with the scattered light field

From a more quantitative analysis, we deduced that this field was generated by the photodiodes of the homodyne detector on EQB1. Using Eq. 6.18, from the spectrum of the homodyne in Fig. 8.1, we can deduce that the scattered light recoupling factor is $f_{sc} \sim 10^{-10}$. This value is compatible with the computation carried out using Eq. 6.7 assuming that the beam is scattered by the photodiodes (BRDF = 50 ppm, $w(z) = 15\,\mu$m and $\alpha_{FI} = 10^3$, the attenuation of the EQB1 Faraday Isolator). The scheme on the left side of Fig. 8.1 shows the LO beam in blue being reflected by the homodyne photodiodes. The scattered field in light blue propagates until the FCIM and then it is reflected and detected by the homodyne detector.

It has been observed that when the cavity is close to resonance, the effect of scattered light is amplified between 20 and 200 Hz, making it impossible to perform squeezing ellipse rotation measurements at frequencies lower than a few hundred Hz. This effect is shown on the spectrum on the right bottom side of Fig. 8.2. The different curves represent the spectra for different detunings of the filter cavity, from resonance condition till 300 Hz detuning. The sharp structures are due to some modulation lines applied on the last two mirrors on the green path on EQB1, called *dither lines*. These are used for the beam pointing control loop used to align the cavity when it is locked with the green beam. The frequencies of the dither lines are 45, 60, 75 and 90 Hz. The EQB1 mirrors angularly displace the green beam arriving on the cavity. However, there is a longitudinal coupling due to the not perfect mechanics of the piezo and to the off-centered beam on the mirrors. This results in an additional longitudinal correction for the longitudinal control loop of

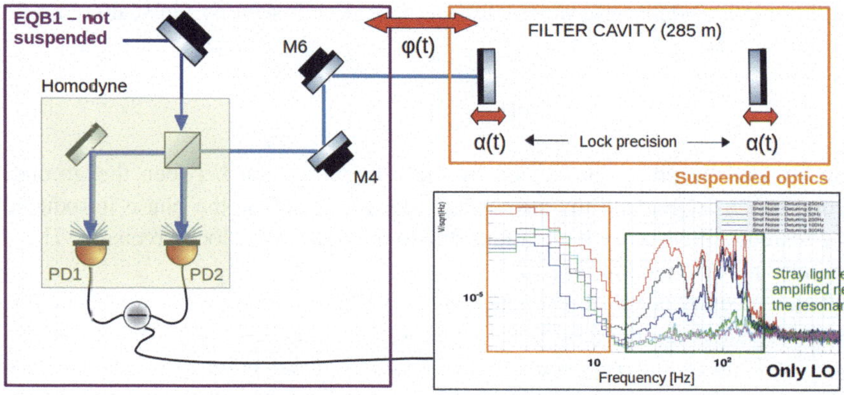

Fig. 8.2 Simplified scheme of the effect of the scattered light from the homodyne photodiodes (in light blue) propagating towards the cavity close to resonance, for different cavity detunings, in different colours on the lower-right plot: 0 Hz (in red), 50 Hz (in black), 100 Hz (in blue), 200 Hz (in green), 250 Hz (in light blue) and 300 Hz (in purple). The effect is amplified between 20 and 200 Hz due to the dither lines on EQB1 mirrors and the low lock precision of the filter cavity. The residual length fluctuation of the filter cavity is introducing an additional phase modulation, $\alpha(t)$

the filter cavity, which at the time of the measurement only acted on the end mirror. However, the infrared beam does not suffer from the disturbance introduced by the dither lines as they are only on the green path. Therefore, the correction of these lines by the longitudinal lock loop introduces noise at the frequencies of the dither lines on the infrared beam. Moreover, this effect is amplified by the low lock precision of the filter cavity as it will be explained in Sect. 8.2. This residual length fluctuation of the filter cavity is seen by the scattered field as an additional phase modulation $\alpha(t)$ (see Sect. 8.2).

8.2 Scattered Light Amplification Due to Filter Cavity Lock Precision

When the cavity is out of resonance the scattered field goes to the input mirror of the filter cavity where it is reflected towards the homodyne. When the cavity is on resonance or close to resonance, the scattered field enters the cavity and it experiences an additional phase modulation. The scattered field phase noise is therefore amplified by the filter cavity.

The principle of scattered light field detection with balanced homodyne was described in Sect. 6.2. The quantity detected by the differential homodyne channel is given by Eq. 6.17. In our case the relative phase between the LO field and the scattered field is

$$\phi_r(t) = \phi_{LO} + \Delta\Phi(t). \tag{8.1}$$

The time-dependent term modulates the optical path seen by the scattered field and is defined as

$$\Delta\Phi(t) = \phi(t) + \alpha(t) \tag{8.2}$$

where $\phi(t)$ is the phase introduced by the relative motion between the ground-attached bench EQB1 and the suspended optics and $\alpha(t)$ is the phase introduced by the residual filter cavity fluctuations due to the not perfect lock precision. These two terms are illustrated in Fig. 8.2.

When the cavity is out of resonance the only phase seen by the scattered field is the one introduced by the motion between EQB1 and FCIM, $\phi(t)$. This phase can be derived from the interference fringes measured on the homodyne detector (see Fig. 8.1), with the cavity out of resonance. The maximum relative motion between the bench ground-attached EQB1 and the suspended optics corresponds to $z_{max} = 10 - 15\,\mu$m. The phase encoding this relative motion, from Eq. 6.1, is therefore

$$\phi_{max} = \frac{4\pi}{\lambda} \cdot z_{max} \sim 4\pi \times 10. \tag{8.3}$$

When the cavity resonates or is close to resonance, this phase shift takes into account also the second term of Eq. 8.2. The phase $\alpha(t)$ can be estimated as

$$\alpha(t) = \frac{4\pi}{\lambda} \cdot \frac{2}{\pi} \cdot F \cdot \delta l(t) \tag{8.4}$$

where $\frac{2}{\pi} \cdot F$ is the number of round-trips that the scattered light does inside the filter cavity ($F = 11{,}000$ is the filter cavity finesse) and δl is the residual relative motion between the mirrors of the cavity, given by the lock precision. The residual motion of the cavity mirrors can be derived from the relation

$$\frac{\delta l_{rms}}{l} = \frac{\delta \nu_{rms}}{\nu} = \frac{15\,\text{Hz}}{3.10^{14}\,\text{Hz}} \tag{8.5}$$

where $l \sim 300$ m is the cavity length, $\delta\nu_{rms} = 15$ Hz is the measured lock precision of the cavity for the infrared light and ν is the laser frequency (1064 nm for the infrared). One gets $\delta l_{rms} = 1.5 \cdot 10^{-11}$ m and therefore, from Eq. 8.4, one obtains

$$\alpha_{rms} \sim 4\pi \times 10. \tag{8.6}$$

It can be seen that the values α_{rms} and ϕ_{max} are of the same order of magnitude. This means that the contributions coming from the relative motion of the ground-attached system and the suspended optics and the one coming from the cavity mirror relative fluctuations provide the same level of scattered light noise.

In order to reduce this noise, it is necessary to improve the filter cavity lock precision δl_{rms} and to compensate the phase $\phi(t)$ seen by the scattered light field. due to the relative motion.

8.3 Stray Light Loop Design

Even if this problem will not be present in Science Mode since the homodyne detector will not be used and the LO can be turned off, it prevents from characterizing of the squeezing rotation at a few tens of Hz before the injection in the interferometer.

In order to eliminate scattered light noise, it is possible to act on two different observables: the amount of scattered light re-coupled in the homodyne f_{sc} and the relative motion between the system and the scattering object $z_r(t)$, as shown in Eq. 6.5.

The first mitigation actions were aimed at reducing the value of f_{sc}. At first we tuned the half wave plate inside the EQB1 Faraday Isolator (see Fig. 4.6). After that we performed a fine tuning of HD photodiodes alignment. Then we installed two irises and a quarter wave plate in front of HD photodiodes to stop back-reflections that could contribute. Finally, we increase the size $w(z)$ of the beam arriving on HD photodiodes in order to reduce the factor f_{sc}, as suggested by Eq. 6.7. The PDs were installed on the waist of the beam. Therefore, we displaced the PDs along the beam direction by few tens of μm, to increase the size on their sensitive area. All these indirect mitigation actions didn't bring any improvement. An other option would be to change both the telescope and the photodiodes to allow for a considerably larger beam size. As this would have been a major modification with a big impact on the planning, we decided not to do this during the commissioning of the FDS system. Finally, it was decided that a Faraday Isolator will be installed in the path of the scattered light on EQB1 to mitigate the problem, while inducing $\sim$2% additional optical losses.

The second actions were aimed at reducing the relative motion between FCIM and the scatterer, i.e. the HD photodiodes on EQB1. In order to do that we implemented a feedback control loop: the Stray Light Loop (SLL). The idea is to actively compensate for the optical path difference seen by the scattered light, which results in the phase $\Delta\Phi(t)$. This is performed by acting on the optical path by means of a piezo-electric actuator that actively changes the optical path of the propagating beam (see Fig. 8.3).

The major difficulty encountered during the implementation of this feedback loop was the interplay of the SLL with the CC loop (described in Sect. 5.4.2). Actually, the two loops act on the same observable, i.e. the optical path difference between the LO and the squeezing field to be detected. Therefore, we ensured that the two loops did not work in conflict but that the action of the SLL reduced the correction sent to the actuators of the CC loop. We designed the SLL so that its Unitary Gain Frequency (UGF) was lower than that of the CC loop (UGF $\sim$ 1 kHz) so that it acts more slowly.

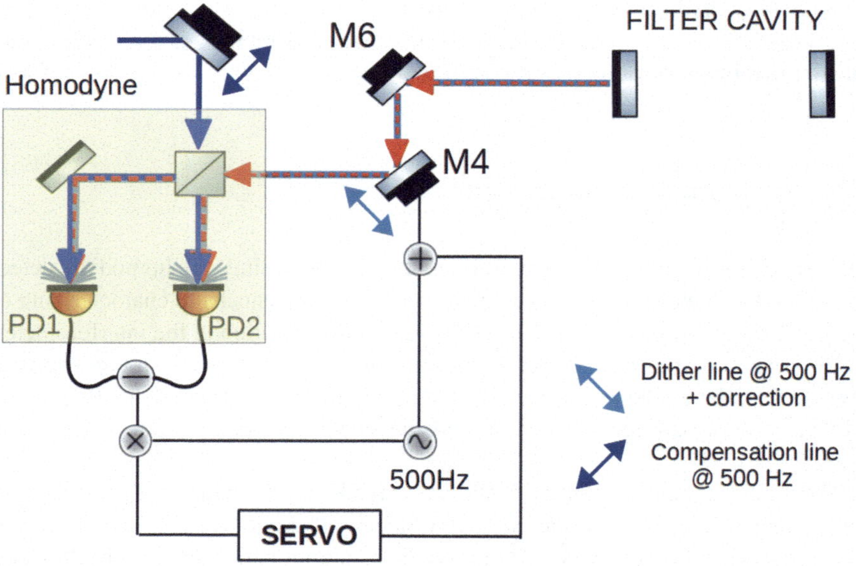

Fig. 8.3 Feedback control loop diagram to mitigate scattered light from the homodyne photodiodes. A 500 Hz dither line is introduced on a mirror in the path of the scattered light (M4). The differential audio signal from the homodyne is demodulated at the dither line frequency to obtain the error signal which is filtered and sent as a correction to the piezo actuator on M4 (and also M6 in the case of high seismic activity). A 500 Hz correction line is introduced on the LO path to compensate for the additional phase noise due to the dither line

The scheme of the feedback control loop implemented on the system is shown in Fig. 8.3. The plant of the system is the low frequency bump detected by the differential audio channel of the homodyne, i.e. the phase difference between the LO and the scattered light fields introduced by the relative motion of the suspended and ground-attached parts of the FDS system. To generate a good error signal, a dither line is applied to the mirror M4 on the path of the scattered light (which coincides with the path of the squeezing reflected by the filter cavity). The error signal is obtained by demodulating the audio differential signal of the homodyne at the frequency of the dither line. The filter applied to the error signal has been optimised to correct with a high gain at low frequencies, introducing what we call a *boost* (see Sect. 8.4.4). The obtained correction is sent to the piezo-electric actuator on the mirror M4 along the beam direction (and in case of high seismic motions also on the mirror M6). In this way the corrected observable is the optical path. Those actuators have a dynamic range of 30 μm each and a bandwidth of $\sim$100 Hz. It is also necessary to introduce a compensation line on the LO path to compensate for the effect of the dither line and avoid introducing additional phase noise in squeezing measurements, as it will be described in Sect. 8.4.3. The compensation line is added on the piezo of the mirror LO_M1 or LO_M2, which have a dynamic range of 3 and 5 μm respectively.

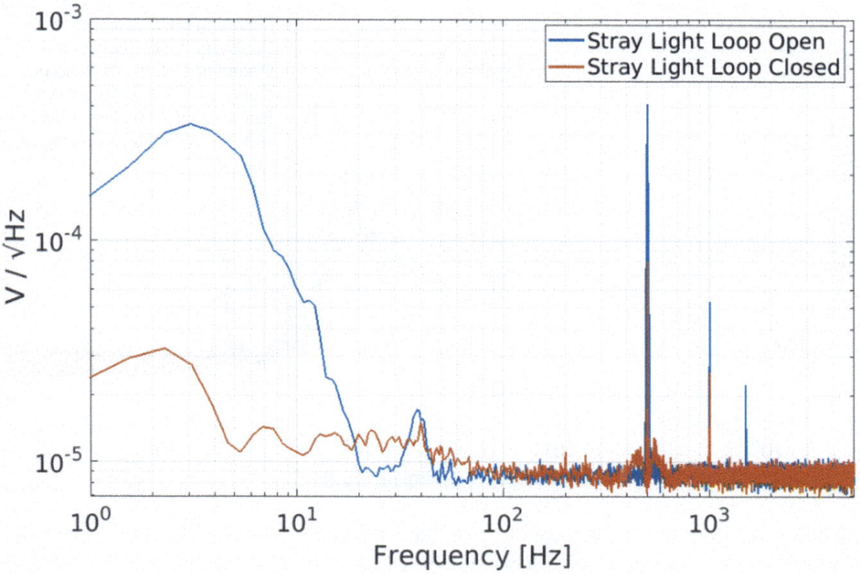

Fig. 8.4 Differential audio channel of the EQB1 homodyne detector with the SLL open (blue) and closed (orange) when the cavity is far from resonance

8.3.1 Stray Light Loop Closed: Preliminary Results

The SLL loop was initially closed by sending only the LO on the homodyne and with a lock precision on the green of about 10–15 Hz. We tested the behavior of the loop for two particular cavity conditions, both far and near its resonance condition. In Sect. 8.2 we saw how the effect of scattered light introduces additional structures when the cavity is close to the resonance.

Figure 8.4 shows the differential audio channel of the homodyne detector when the SLL is open (blue) and closed (orange), when the cavity is far from the resonance. The error signal is filtered only with an integrator, see Sect. 2.4.1. The bump at low frequencies is reduced by an order of magnitude thanks to the SLL. However, the noise is reintroduced between 20 and 100 Hz. This means that in this configuration it would not have been possible to measure squeezing below 100 Hz. In conclusion, the SLL works but it reintroduces the noise at higher frequencies. Therefore, the control filter needs to be optimized, as it will be described in Sect. 8.4.4.

If the filter cavity is at resonance or is detuned by tens or a few hundred Hz, there is still the noise explained in Sect. 8.2, as shown in Fig. 8.5. To solve this problem, the lock precision of the filter cavity was improved and the dither lines for FC alignment on the green mirrors were removed to reduce the effect of scattered light between 20 and 200 Hz near the FC resonance condition.

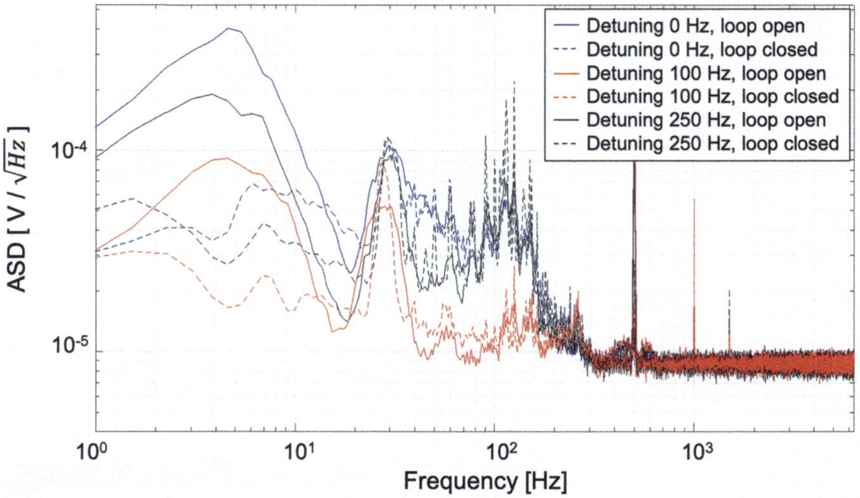

Fig. 8.5 Amplitude spectral density of the differential audio channel of the EQB1 homodyne detector with the SLL open (solid lines) and closed (dashed lines) for different FC detunings: 0 Hz (blue), 100 Hz (red) and 250 Hz (black)

8.4 Stray Light Loop Optimization

In order to maximize the active scattered light reduction, we characterized and optimized the loop. We measured its Transfer Function, optimized the dither line amplitude, compensated the 500 Hz line acting on the LO path and tuned the correction filter.

8.4.1 Stray Light Loop Transfer Function

To characterize the loop, it is necessary to measure its UGF (see Sect. 2.4.1) and making sure that it is lower than the CC loop one (∼3 kHz, see Sect. 5.4.2) to avoid the loops to be in conflict. The open loop transfer function has been measured with the SLL closed, sending only the LO on the homodyne detector to be in a simple configuration. It has been measured from Eq. 2.58 being $Y(s)$ the perturbation sent on the mirror M4 (along the direction of the beam Z) and $X(s)$ the signal detected by the homodyne detector. We injected a sine wave noise at different frequencies (*swept sine* injection) to measure the amplitude of the line in the transfer function. The graph of the SLL Transfer Function (TF) amplitude is shown in Fig. 8.6, taking the TF amplitude for every sine frequency injection. The UGF of the loop is at ∼12 Hz. This is fine and it is of the order of magnitude that we expect in order not to be in conflict with the CC loop.

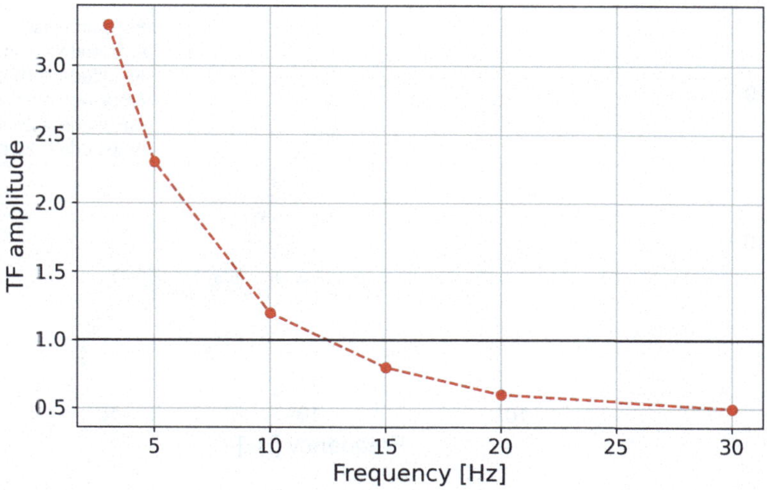

Fig. 8.6 Transfer Function amplitude of the SLL computed from sine wave noise injections. The UGF of the loop is at ~12 Hz

8.4.2 Dither Line Amplitude Optimization

The amplitude of the 500 Hz dither line sent to the M4 mirror has to be optimized in order to maximize the Signal-to-Noise ratio (SNR) of the error signal. It has been done with the SLL closed sending only the LO beam on the homodyne. The SNR is calculated as the ratio between the error signal level at low frequency and the sensing noise, which corresponds to the signal in the absence of the dither line. The sensing noise and error signal for different dither line amplitudes are shown in Fig. 8.7. For amplitudes higher than 15 mV, the SNR doesn't improve significantly. The optimal amplitude in this configuration of the system is therefore 15 mV that corresponds to a displacement of 0.13 μm and to a SNR ~ 10.

8.4.3 Compensation of the Dither Line

In order to avoid to introduce additional phase noise to the squeezing detected by the homodyne, the dither line applied on M4 must be compensated. The effect of phase noise can be seen from the beatnote signal between the CC field and the LO beam. Since with the CC loop open this noise is too high, it is convenient to close this loop to stabilize the phase and keep the error signal in its linear region. To compensate for the 500 Hz line which is introduced on the squeezing/CC path, a modulation at the same frequency must be introduced in the LO optical path. The optimization is done by minimizing the 500 Hz line in the CC error signal. The parameters to be tuned are the amplitude and the initial phase of the compensating

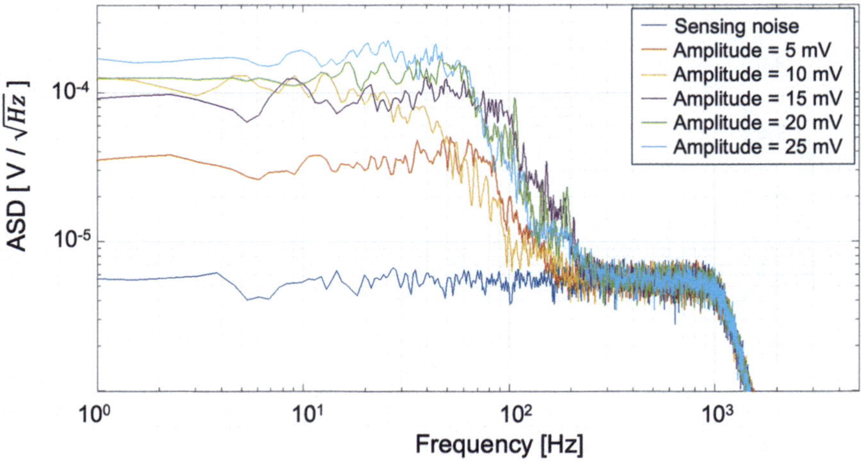

Fig. 8.7 Optimization of the dither line amplitude in order to maximize the Signal-to-Noise Ratio (SNR) of the error signal of the SLL

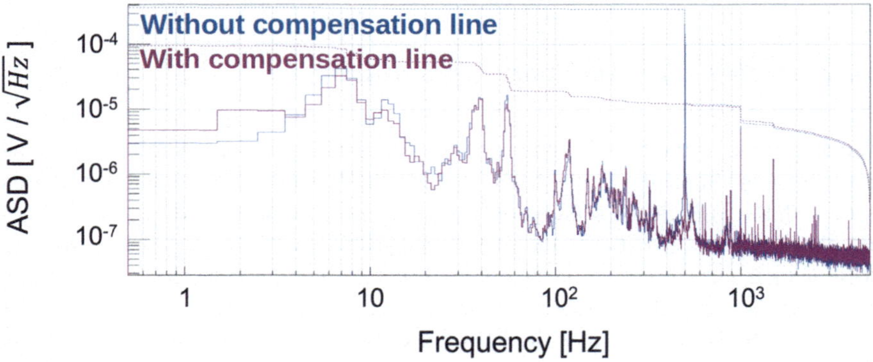

Fig. 8.8 Optimization of the compensation line on the CC loop error signal. The blue solid curves represents the signal without the compensation line and the blue dashed line its RMS. The magenta solid curve represents the signal with the optimized compensation line and the magenta dashed line its RMS

line. The amplitude can be derived by inter-calibrating the two actuators in question, the one on M4 to which the dither line is applied and the one on LO_M2 to which the compensation line is applied. For a dither line of amplitude $A_d = 15\,\text{mV}$, we found that the optimal compensation line amplitude is $A_c = 100\,\text{mV}$. The comparison between the CC signal with and without compensation line is shown in Fig. 8.8, in magenta and blue respectively. Despite the 500 Hz dither line is compensated, its harmonics (at 1000 and 1500 Hz) are very high and they are worsening the phase noise. It is planned to compensate the harmonics sending compensating lines on other two CC actuators on the LO path.

8.4.4 Correction Filter Optimization

It is needed to find a filter which allows to reduce the noise at low frequency and does not add noise at high frequency. A simple low pass filter, i.e. an integrator (see Sect. 2.4.1), it is not the optimal solution since it does not suppress all the low frequency noise. It is therefore needed a more complex filter.

The dither line sent on M4 has an amplitude of 15 mV and the compensation line sent on LO_M2 has an amplitude of 100 mV, as described above. We optimized the gain of the loop at the beginning, using as filter an integrator. The optimal gain value corresponds to the higher bump attenuation without re-injecting noise at higher frequencies. Then we added a pair of simple pole and zero (see Sect. 2.4.1) at different frequencies to able to send a larger correction at low frequency and to filter the high frequency, that is what we refer to as *boost*. The different configurations tried are shown in Fig. 8.9 and they are

- SLL open, integrator (pink);
- SLL closed, integrator, gain = 500 (yellow);
- SLL closed, boost: pole at 0.01, Q = 0 and zero at 10 Hz, Q = 0, gain = 2500 (light green);
- SLL closed, boost: pole at 0.01, Q = 0 and zero at 7 Hz, Q = 0, gain = 2500 (green);

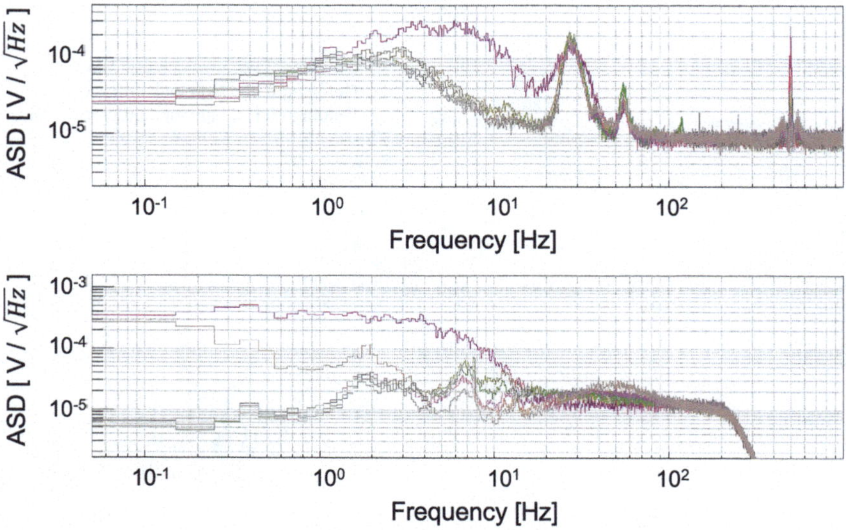

Fig. 8.9 Differential audio channel of the homodyne on the top and SLL error signal channel on the bottom. The different curves represent different gains and parameters for the correction filter (as listed in the text). The optimal configuration is filtering with a boost at low frequency with a pair pole (f = 0.1, Q = 0) and zero (f = 5 Hz, Q = 0) and gain = 2500 (light pink curve)

- SLL closed, boost: pole at 0.01, Q = 0 and zero at 5 Hz, Q = 0, gain = 2500 (light pink);
- SLL closed, boost: pole at 0.01, Q = 0 and zero at 3 Hz, Q = 0, gain = 2500 (grey).

The best configuration is with the boost: pole at 0.01, Q = 0 and zero at 5 Hz, Q = 0, gain = 2500 (light pink), since it introduces less noise at high frequency (as seen in the error signal).

8.4.5 Stray Light Loop Closed with High Lock Precision

As described in Sect. 8.2, the lock precision of the filter cavity introduces additional scattered light noise. Therefore the longitudinal control of cavity with the green beam has been improved by sending the correction of the locking loop on both the end mirror of the cavity and the frequency of the laser. The lock precision is increased in this way from 10 Hz to about 1 Hz, as described in Sect. 5.6.1. The loop was closed in this condition by placing the cavity at its resonance or slightly detuned. The effect of the new longitudinal cavity control loop and the SLL closed in this condition, is shown in Figs. 8.10 and 8.11 (for the filter cavity in resonance and detuned by 100 Hz respectively). The blue curves represent the effect of scattered light with low lock precision and the red ones with improved lock precision (∼2 Hz lock precision). The structures present between 20 and 200 Hz are reduced when the

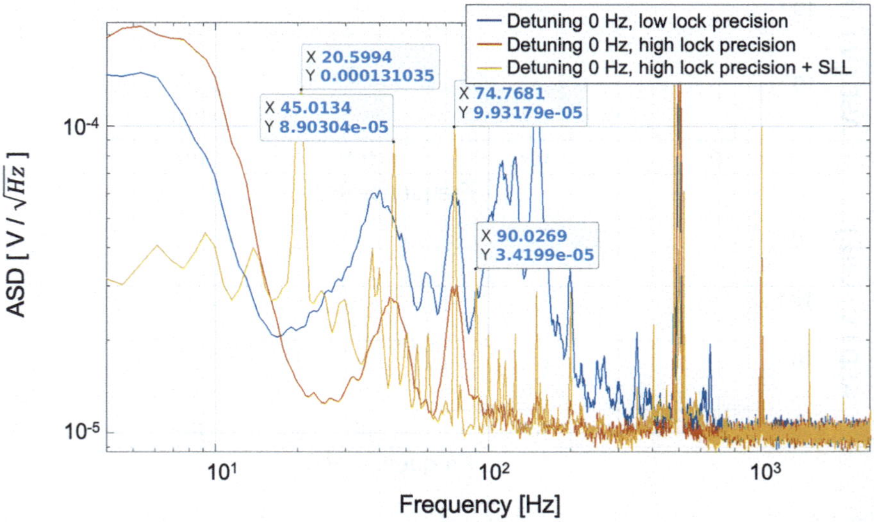

Fig. 8.10 Homodyne differential audio channel only with the LO beam in low lock precision condition (in blue) and in high lock precision condition (in red) with the filter cavity on resonance. In yellow the behavior of the scattered light with the SLL closed in high lock precision condition. The peak at 20 Hz is an oscillation of the loop that was not optimized during the measurement

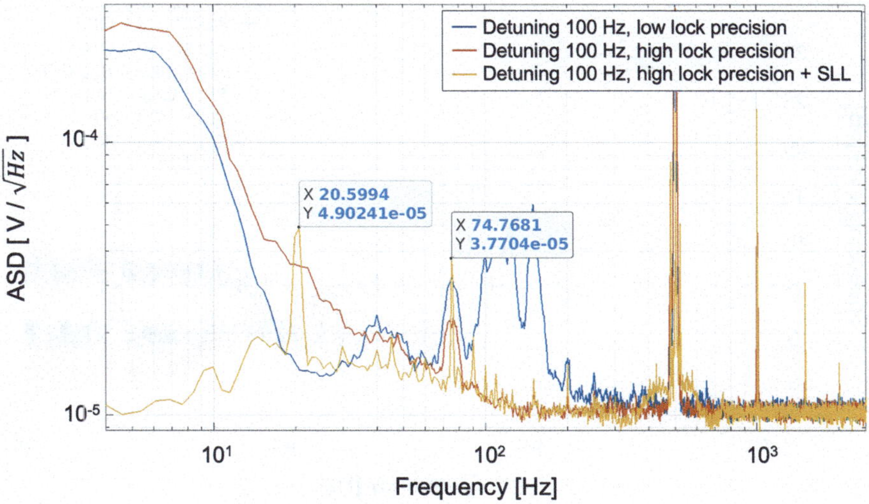

Fig. 8.11 Homodyne differential audio channel only with the LO beam in low lock precision condition (in blue) and in high lock precision condition (in red) with the filter cavity detuned by 100 Hz. In yellow the behavior of the scattered light with the SLL closed in high lock precision condition. The peak at 20 Hz is an oscillation of the loop that was not optimized during the measurement

lock precision improves because the residual longitudinal motion δl_{rms} of the cavity decreases (see Sect. 8.2). The yellow curves refer to when the scattered light loop is closed with high lock precision. The structures due to the scattered light noise above the UGF ($\sim$12 Hz, see Sect. 8.4.1) are not corrected by the SLL. The loop however reduces the optical path oscillations introduced by the relative motion $z_r(t)$. Therefore the scattered light noise at the dither lines frequency (the ones applied to the mirrors on the EQB1 green path) is unchanged but closing SLL narrows the noise around the dither frequencies. This happens because the modulation of the optical path is lower, see Eq. 6.14. Since the amount of scattered light does not vary, the area subtended by the noise peaks is the same, but with the loop closed, the peaks are sharper.

The next step is to change the cavity angular control strategy in order to avoid the effect introduced by the dither lines on the green.

8.5 Squeezing Measurements with Stray Light Loop Closed

After optimizing the loop using only the LO beam, the final step is to run the SLL together with the CC so that squeezing measurements can be made without the scattered light noise at low frequency. Since both loops act on the same physical observable, i.e. the relative optical path between the squeezing beam and the LO

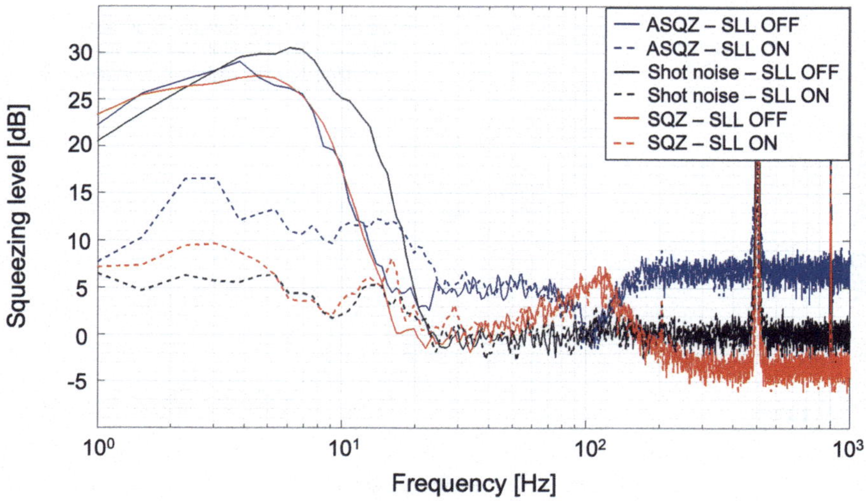

Fig. 8.12 Frequency dependent squeezing measurement at 120 Hz with the SLL loop both open (continuous lines) and closed (dashed lines), shot noise in black, squeezing in red and anti-squeezing in blue

beam, the direction of the loops has been set to make them agree by acting on the gains of the two loops. We tuned the filter of the SLL to optimize the suppression of the scattered light bump in squeezing mode and with the CC loop engaged. The final SLL filter parameters are: double pole at 0 Hz, Q = 0; pole at 0.01 Hz, Q = 0; zero at 0.5 Hz, Q = 0; pole at 15 Hz, Q = 0; gain at 1 Hz of 250. In Fig. 8.12 the homodyne audio differential channel during the squeezing (in red), shot noise (in black) and anti-squeezing (in blue) measurements is shown with the cavity detuned by 120 Hz. The solid lines refer to when the SLL is open and the dashed lines to when SLL is closed. It can be seen that it is possible to run the two loops together and that no noise is reintroduced between 20 and 200 Hz thanks to the high filter cavity longitudinal lock precision (1 Hz). Furthermore, the structures seen in Sect. 8.4.5 are not present as the angular control of the cavity no longer uses dither lines but the wavefront sensing technique, performed with quadrants in reflection from the cavity. The scattered light noise between 5 and 8 Hz is reduced by 25 dB for squeezing and shot noise. However, this is still not the ideal loop configuration as a further 5 dB reduction is needed to suppress all the noise.

Figure 8.13 shows the behaviour of the loop with a detuning of 50 Hz, which is close to the target of AdV+. The level of anti-squeezing goes below the shot noise at low frequencies but it is necessary to better optimize the SLL loop so that the reduction of the noise can be appreciated from 20 Hz downwards.

However, it was not possible to go any further in optimizing the loop due to a hardware problem in our system. Indeed, the actuator range of the coherent control is not sufficient to guarantee a stable CC loop behavior during the measurements. This prevents from being able to fine-tune the SLL filter and compensate the dither line at 500 Hz. Actually, the CC coarse corrections on the LO_M1 and

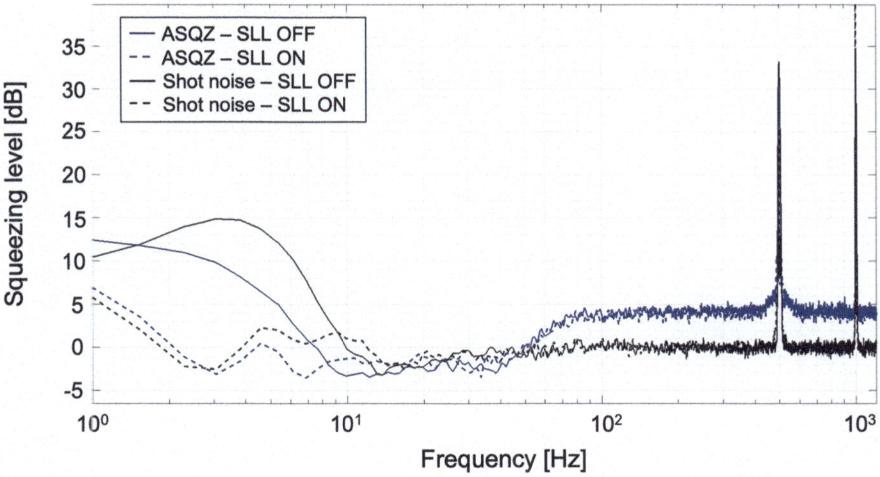

Fig. 8.13 Frequency dependent squeezing measurement at 50 Hz with the SLL loop both open (continuous lines) and closed (dashed lines): shot noise in black and anti-squeezing in blue

LO_M2 actuators diverge periodically (every ∼40 seconds, depending also on seismic motion conditions), without unlocking the CC loop, but making the SLL unstable. The solution to this problem is to replace these low dynamic range actuators (∼3–5 μm) with a higher dynamic range actuator such as the one used on M4 (∼30 μm).

8.6 Conclusion

In this chapter is reported an evidence of scattered light on the frequency-dependent squeezing system, generated by the homodyne detector photodiodes. This noise, even if it will not be present in Science Mode, does not allow to characterize the rotation of the squeezing at low frequencies. In this chapter it is described a feedback control loop implemented to correct the phase variation introduced by the relative motion between the scatterer and the system. The control loop has been successfully tested and optimized. Nevertheless, it is supposed to work at the same time as the loop fixing the detection angle of the squeezing (the coherent control loop). In order to finish this optimization it is needed to replace a piezo-electric actuator with one with higher dynamic range. The next step will be to substitute the actuator and tune the performances of the loop in order to be able to fully correct the scattered light noise. In this way it will be possible to characterize the frequency-dependent squeezing before injecting it in the interferometer, as it will be described in the next chapter.

Part IV
Preparation for Squeezing Injection into the Interferometer

Chapter 9
Matching of the Squeezing Beam to the Interferometer

The squeezed beam must be well matched with the interferometer beam, i.e. it must be aligned and mode matched. A misalignment and mode mismatch (see Appendix A) between the two beams worsen the squeezing performance in the form of optical losses as described in Sect. 3.4.1. The optical losses introduced in this way must remain below 1% [74].

In order to be able to estimate the mode matching, it is necessary to measure the parameters of the interferometer beam and of the squeezing beam before it is injected. Then, the matching is improved by adapting the telescope on the squeezing path in order to properly match it to the ITF beam.

In this chapter we describe how the mode matching between the squeezing beam and the interferometer beam has been measured and optimized.

9.1 ITF-FDS System Interface

The squeezing produced on the FDS system (both FIS and FDS) needs to be injected into the interferometer through the anti-symmetric port of the ITF. The interface between the two systems is composed of optics on the SDB1 and SQB1 benches, which are connected by a minilink, as shown in Fig. 9.1.

The squeezed beam is sent from SQB1 into the Faraday Isolator on SDB1 and then towards the interferometer, as shown by the purple arrows.

The pre-alignment and pre-mode matching of the squeezed beam sent towards SDB1 were performed at LAPP, simulating the interferometer beam through a telescope, as described in Sect. 5.1. However, the system alignment and matching need to be verified and optimized on site where the actual parameters of the squeezed beam and the interferometer beam can be directly measured. The latter is the spurious beam indicated by the pink arrows in Fig. 9.1. It is the p-polarized light reflected by the Faraday Isolator on SDB1, having very similar parameters

E. Polini, *Broadband Quantum Noise Reduction in Advanced Virgo Plus*,
Springer Theses, https://doi.org/10.1007/978-3-031-95143-5_9

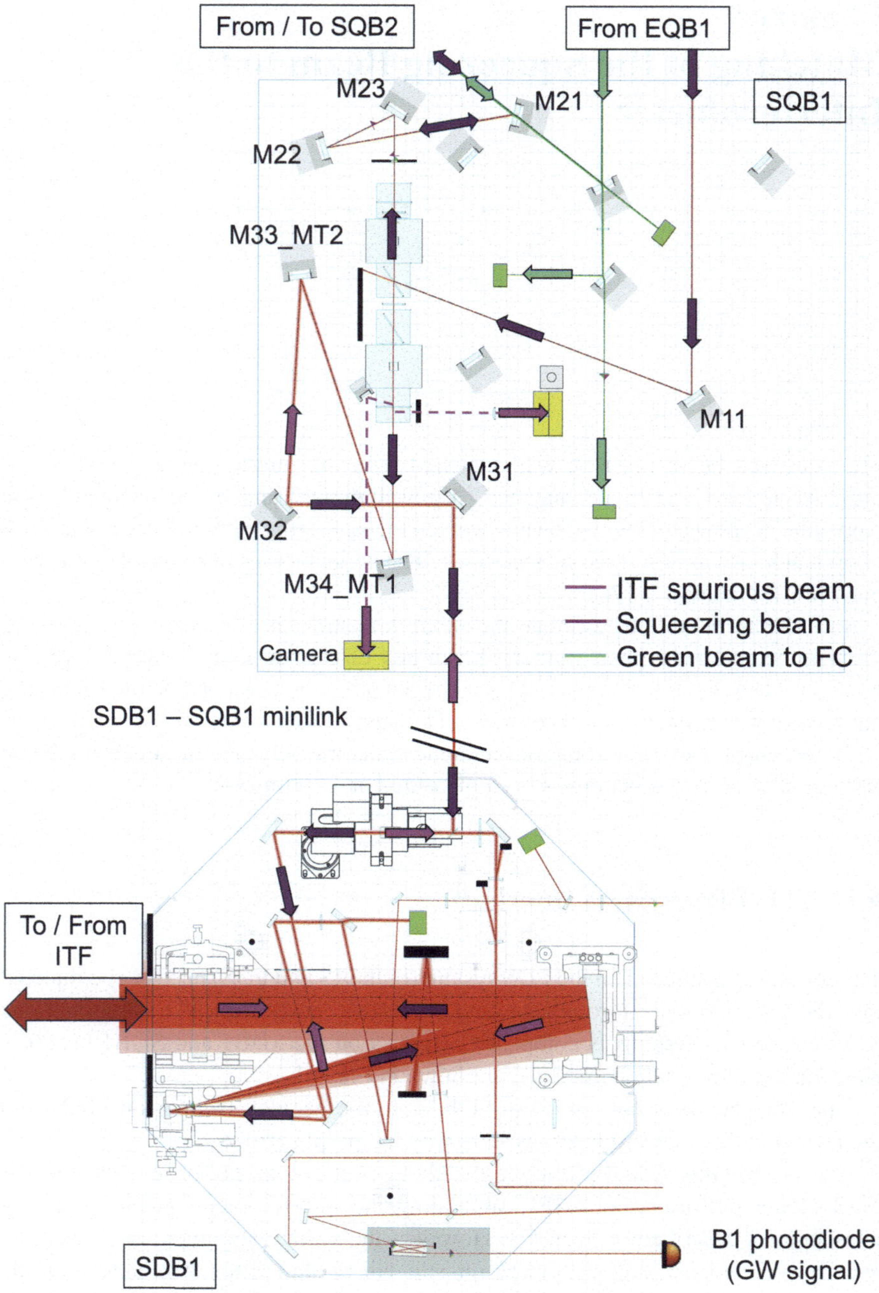

Fig. 9.1 Interface between the squeezing system (SQB1 on the top) and the detection system (SDB1 on the bottom)

as the s-polarized dark fringe beam. The initial work is conducted with the SQB1 bench in-air in order to measure the parameters of the ITF and squeezed beams with a beam profiler. The pre-matching has been performed by adapting the telescope on SQB1 composed by the mirrors M33 and M34. Once the parameters of the ITF beam and the squeezed beam sent to SDB1 are considered close enough, an alignment loop will be implemented using the interferometer spurious beam arriving on SQB1. This beam is measured by the cameras on SQB1 (one in far field and the other in near field). In order to keep the beams on the desired positions on the cameras, a correction is sent to the tip-tilt PZT mirrors M31 and M34, on SQB1 to keep the beam on the desired position on the cameras. The squeezed beam propagates inside the interferometer and then it returns on SDB1 where it passes into the output mode cleaner cavity (OMC), see Chap. 10. The fine alignment it is done by looking at the transmission of this cavity. Since the squeezing beam is not visible, the BAB which has the same parameters (see Sect. 4.2.3) will be used for this purpose.

9.2 Squeezed Beam Alignment and Mode-Matching

Once the FDS system was commissioned and well aligned (see Sect. 5.3), we measured the parameters of the beam coming from the FC on SQB1 to check if it was in agreement with what was expected. The beam propagation study was carried out with the BAB, which has the same characteristics (beam waist and Rayleigh range) as the squeezed beam.

We started by measuring the size of the beam reflected from the filter cavity and directed towards EQB1. We positioned a beam profiler at different positions and from the Gaussian fit of these data we determined the beam parameters as explained in Sect. 5.1.1. The parameters from the fit of the data measured between M13 and M14 on SQB1, are:

- $w_{0x} = 590\,\mu\text{m}$ located at 1081 mm from M13;
- $w_{0y} = 660\,\mu\text{m}$ located at 1155 mm from M13;

and are in agreement with the expected design values:

- $w_{0x} = 640\,\mu\text{m}$ located at 1211 mm from M13;
- $w_{0y} = 720\,\mu\text{m}$ located at 1461 mm from M13.

The beam waist is located close to M23 mirror.

Afterwards, we tuned the second waveplate inside the double FI on SQB1 in order to transmit the beam in the direction of the interferometer. Then we diverted the beam after the last mirror on the bench, SQB1_M31, with temporary mirrors to measure the beam profile, as shown on the left of Fig. 9.2.

The expected values for the beam going to the interferometer must be the same as the ones of the beam coming from the interferometer, since they must be matched. From the simulation, the expected parameters of the two beams are

- $w_{0x} = w_{0y} = 1350\,\mu\text{m}$ located at ± 3650 mm from M31

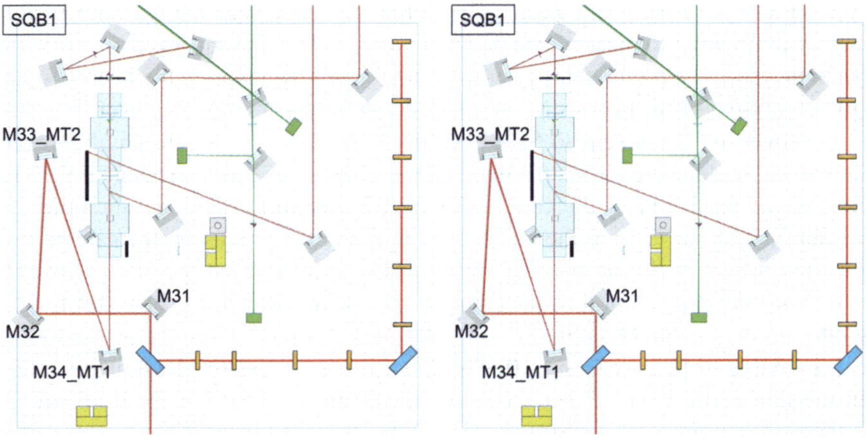

Fig. 9.2 Positions of the beam profiler during the data taking. On the left is shown the measurement of the beam reflected by the filter cavity going to the ITF; on the right is shown the measurement of the spurious beam coming from the ITF

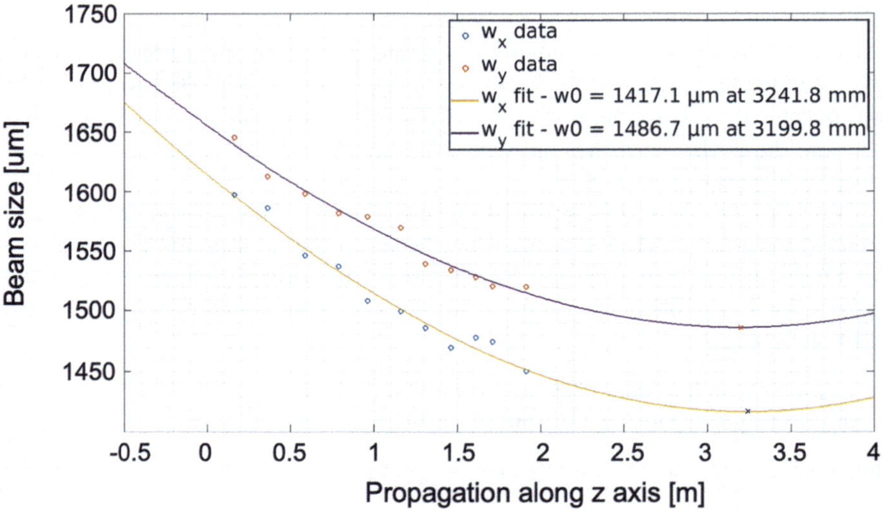

Fig. 9.3 Profile of the BAB beam in the positions shown in Fig. 9.2 (left), i.e. from the FC to the ITF. The position zero refers to the SQB1_M31 mirror position

where the sign of the distance is positive for the squeezed beam and negative for the ITF beam, due to the different propagation direction during the measurements.

The profile of the beam going to the ITF is shown in Fig. 9.3. The fit gives the following parameters:

- $w_{0x} = 1419\,\mu\text{m}$ located at 3257 mm from M31;
- $w_{0y} = 1482\,\mu\text{m}$ located at 3232 mm from M31.

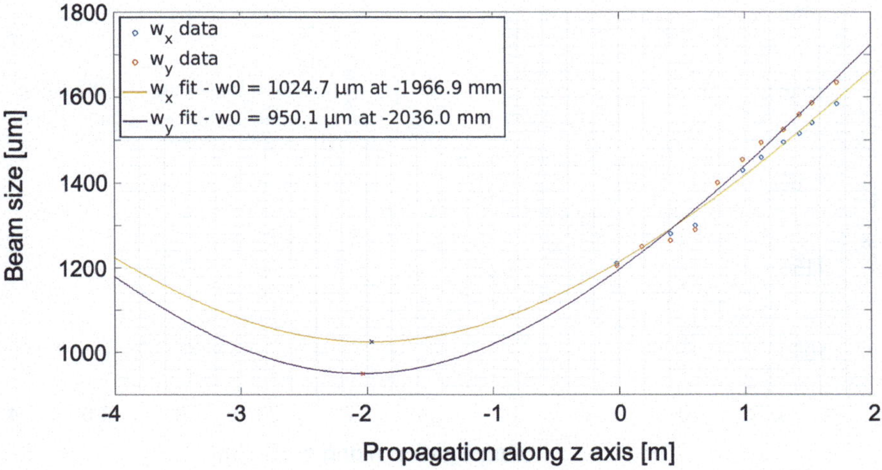

Fig. 9.4 Profile of the ITF spurious beam in the positions shown in Fig. 9.2 (right). The position zero refers to the SQB1_M31 mirror position

The profile of the beam going to the ITF is considered to be close enough to what we expected.

Afterwards, we measured the spurious beam coming from the ITF, placing the beam profiler where indicated on the right of Fig. 9.2. The ITF was set in its simplest configuration, i.e. with the input mirror aligned and all the other mirrors misaligned. This configuration is called *single bounce*. From the data of the beam profile the fit which gives the following parameters, as shown in Fig. 9.4:

- $w_{0x} = 1025\,\mu\text{m}$ located at -1967 mm from M31;
- $w_{0y} = 950\,\mu\text{m}$ located at -2036 mm from M31;

which is very far from the expected values.

However, this value is similar to the beam parameters from the ITF measured in O3, which were [108]:

- $w_{0x} = w_{0y} = 950\,\mu\text{m}$ located at -1600 mm from the steering mirror on the external squeezer bench (similar reference as M31).

The mode matching between the beams coming from and going to the ITF is computed from their overlap integral (see Eq. 6.11) and it is 84%.

Two mirrors of the telescope on SQB1, M33 and M34, are mounted on translation stages (of range ± 1.25 cm) as shown in Fig. 4.11 to match the two beams. However, it is not possible to correct such a high mode mismatch by acting only on the translation stages. According to the simulations performed with Optocad it is necessary to replace one mirror of the telescope, M34, with another one with a different radius of curvature (RoC).

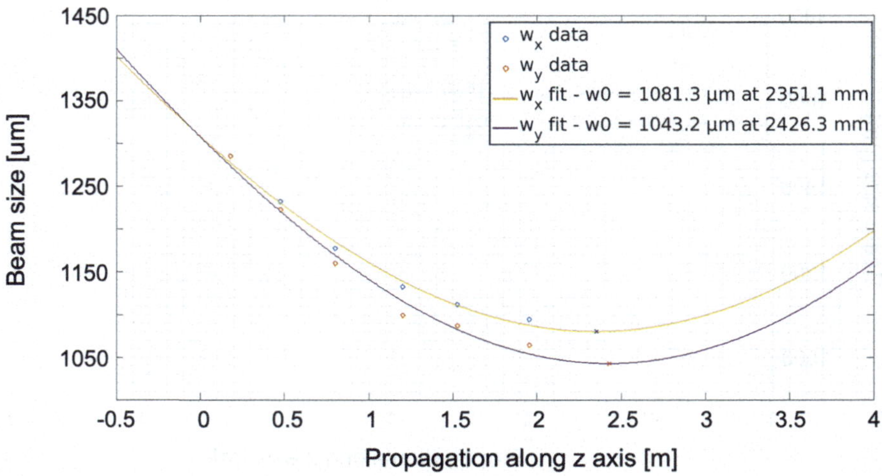

Fig. 9.5 Beam reflected by the FC measured on SQB1 and going to the ITF and Gaussian fit. The position zero refers to the SQB1_M31 mirror position

9.2.1 Telescope Adaptation

The beam coming from the interferometer does not have the expected size and it is not yet understood why. Nevertheless, in order to match the squeezing beam to the ITF beam, it was decided to modify the telescope on SQB1. According to the simulations, the mirror M34, that initially had a RoC = −1.35 m, needs to be replaced by a mirror with RoC = −2.125 m. After replacing this mirror, the beam reflected by the filter cavity and directed towards the interferometer was measured again. The fit of the measured beam profile data indicates that the beam has the following parameters, as shown in Fig. 9.5:

- $w_{0x} =$ 1081.3 μm located at 2351 mm from M31;
- $w_{0y} =$ 1043.2 μm located at 2426 mm from M31.

These values are compatible with those of the beam arriving from the interferometer, which have been re-measured during the same intervention:

- $w_{0x} =$ 920.6 μm located at −1902.4 mm from M31;
- $w_{0y} =$ 1076.8 μm located at −1968.3 mm from M31.

The overlap integral (see Eq. 6.11) between the two beams gives a mode matching between the incoming beam and outgoing beam of $\sim$99.3%. The mode mismatch is therefore $\leq$1% which is the specification for the mismatch losses.

This is a good result that will allow the fine alignment of the BAB in the OMC and the implementation of control loops between ITF and FDS system in the coming months.

9.3 Conclusion

In this chapter the interface between the squeezing system and the interferometer has been described. The matching between the squeezed beam and the interferometer beam has been measured to be 84% which was too low compared to the required level of 99% needed not to limit the losses. We showed that it is necessary to modify the telescope in order to improve the mode matching. After the adaptation of the optical configuration, a mode matching of 99.3% has been measured which is compatible with the expected level of losses. Moreover, the interferometer beam needs to be matched with the cavity which cleans the output beam containing the GW signal, the output mode cleaner cavity. Therefore, the fine mode matching tuning of the squeezing with the ITF beam will be performed by using the transmission from this cavity. More details on the output mode cleaner will be given in the next chapter.

Chapter 10
New High Finesse Output Mode Cleaner for Advanced Virgo Plus

The output mode cleaner (OMC) cavity cleans the beam transmitted at the anti-symmetric interferometer port of all higher-order modes and RF sidebands (introduced for the interferometer control, see Sect. 2.4.2). The replacement of the two low finesse output mode cleaner (OMC) cavities with a high finesse one was one of the major upgrades between the O3 and O4 observation runs. The main motivation for this upgrade is to reduce losses at the output port of the interferometer in order to reduce the quantum noise. In Table 10.1 the losses budget measured during the observing runs O3 and expected for O4 for the detection system is detailed. The total losses should be reduced by a factor 2.

During O3, one of the main limitations was the relative mode mismatch losses between the two mode cleaners. With only one cavity this problem is solved. Moreover, by motorizing the half-waveplates placed before the OMC, it is possible to fine tune the polarization matching between the beam coming from the interferometer and the OMC, when the bench is in vacuum and suspended. After these improvements, detection losses have been measured to be reduced by a factor 2 as will be shown in this chapter.

10.1 Output Mode Cleaner Cavity Design

The OMC is a Fabry-Perot cavity placed at the output port of the ITF with the aim to clean the carrier beam containing the gravitational wave signal, as shown in Fig. 10.1.

The input beam is filtered by the cavity and only the TEM00 mode (see Appendix A), carrying the GW information, is transmitted when the OMC is on resonance. The reflected beam contains instead the higher order modes and the RF sidebands. To design this cavity, several parameters were chosen to meet different requirements: the geometry of the cavity, the radius of curvature of the spherical

E. Polini, *Broadband Quantum Noise Reduction in Advanced Virgo Plus*,
Springer Theses, https://doi.org/10.1007/978-3-031-95143-5_10

Table 10.1 Losses budget of the detection system between the observing runs O3 and O4, both pessimistic and target scenarios. In the pessimistic scenario the two cavities will not be replaced with the single one

DET components	Losses in O3	Pessimistic losses in O4	Target losses in O4
SDB1 Faraday Isolator	1%	1%	0.8%
B1p pick-off	1.5%	0.75%	0.75%
ITF-OMC1 matching	8%	5%	3%
OMC1-OMC2 matching	2.5%	2.5%	0
OMC internal losses	2%	2%	2%
B1 PDs QE	1%	1%	1%
B1 PDs dark noise	4%	4%	1%
B1 minilink viewport	0.04%	0.04%	0.04%
B1 path lenses	0.26%	0.26%	0.26%
B1 path mirrors	0.1%	0.1%	0.1%
Total	20.4%	16.7%	9.0%

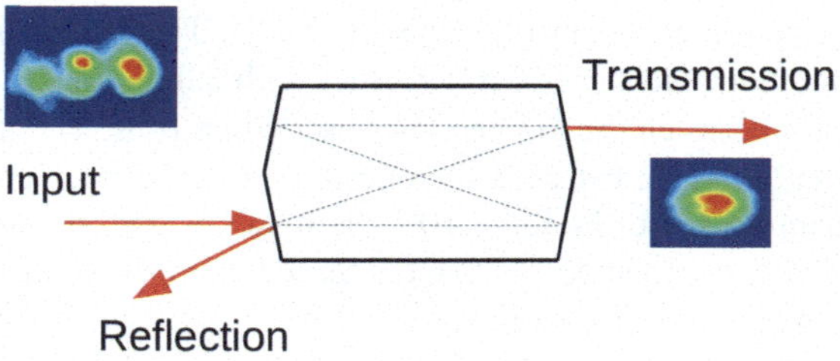

Fig. 10.1 OMC cavity design. The input beam comes from the interferometer and it contains the carrier and the higher order modes. The cavity cleans the beam and transmits the TEM00 that will reach the B1 detector. The reflected beam contains the higher order modes

surface and the finesse. This cavity must be designed to ensure that no noise is introduced that worsens the sensitivity of AdV+. In particular the dominant noises associated to the OMC cavity are the cavity length noise and the thermo-refractive noise.

The cavity geometry parameters and radius of curvature remain the same as for the cavities used in O3 and a more detailed discussion can be found in [109]. The main parameters are listed below, justifying their choice.

10.1.1 Geometry

The Output Mode Cleaner is a bow-tie monolithic cavity in fused silica, Suprasil 3001, as the previous cavities used in Virgo. This material guarantees low absorption and it is very homogeneous. The cavity consists of four polished surfaces, three of which are flat and one of which is spherical, as indicated in Fig. 10.2. The cavity

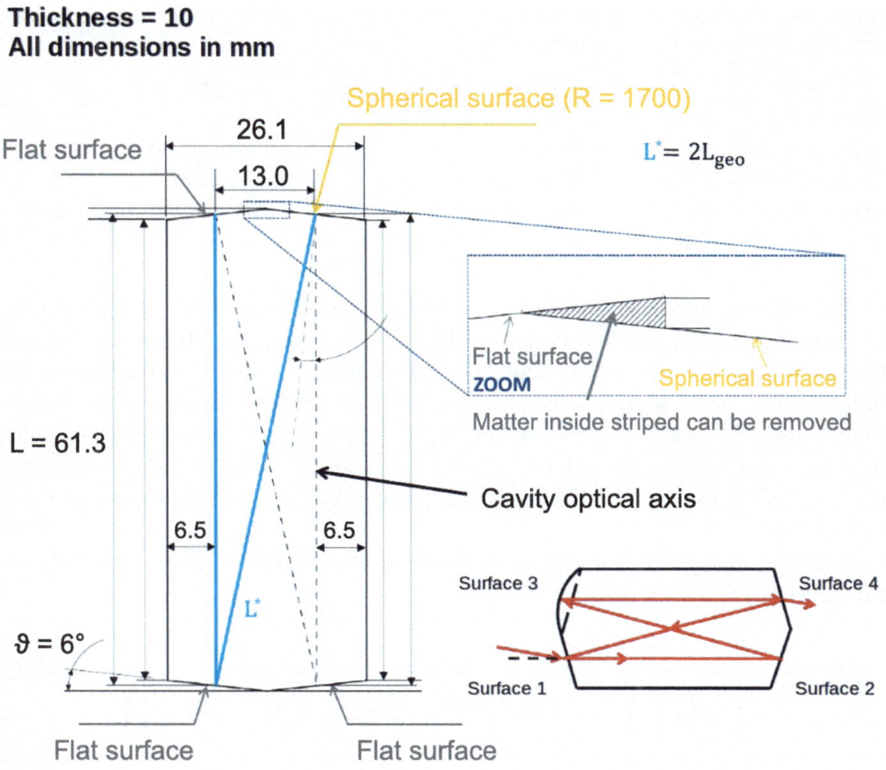

Fig. 10.2 Geometry of the output mode cleaner cavity, after polishing. The dimensions are indicated in mm, and the angles in degrees. The piece thickness is 10 mm. In blue it is indicated $L^* = 2L_{geo}$. Figure adapted from [110]

dimensions after the polishing, are the same as for the cavities used during the last run and they are shown in Fig. 10.2. The specifications on surface polishing quality of each of the four surfaces are:

- The clear aperture diameter is 8 mm;
- The micro-roughness within clear aperture is $\sigma_{RMS} \leq 0.1$ nm RMS.

The dimensions of the cavity are detailed in Fig. 10.2 and one can define:

- The geometrical length as $L_{geo} = \dfrac{L + \sqrt{l^2 + L^2}}{2} = 62 \pm 0.2$ mm;
- The optical length as $L_{opt} = 2nL_{geo}$, where $n = 1.45$ is the refractive index of the fused silica.

In Table 10.2, the comparison of the main characteristics between the cavities used in the past (two identical cavities) and the new OMC cavity that will be used during the observing run O4 is shown.

Table 10.2 Comparison between the main specifications of the OMC cavities used during O3 and the OMC cavity that will be used in O4

Properties	O3 OMC cavities	O4 OMC cavity
Finesse (F)	123	1000
Micro-roughness (σ_{RMS})	0.3 nm RMS	$\leq$0.1 nm RMS
Radius of Curvature (ρ)	1700 mm	1700 mm
Material	Suprasil 3001 (fused silica)	Suprasil 3001 (fused silica)

The internal losses in the cavity are mainly due to the scattering by the four internal surfaces, as detailed in Sect. 10.3.4. These losses are proportional to the finesse of the cavity, since they depend on the number of round-trips the beam does in the cavity. The scattering losses are also proportional to the square of the micro-roughness of the surfaces, as shown in Eq. 10.12. Therefore, to counterbalance the increase of internal losses due to the higher finesse, the value of micro-roughness rms has been improved by a factor 3. This compensate the increase in finesse from 123 to 1000.

The transmission factor for each mode of the cavity is given (similarly to Eq. 2.25) by

$$T(\nu, N) = \frac{1}{1 + \left(\frac{2F}{\pi}\right)^2 \sin^2\left(\frac{2\pi L_{opt}\nu}{c} - N \arccos\left(\sqrt{1 - \frac{2L_{geo}}{\rho}}\right)\right)}. \tag{10.1}$$

where ν is the beam frequency (ν_0 for the carrier, $\nu_0 + \Delta\nu$ with $\Delta\nu = 6,56$ MHz for the RF sidebands) and $N = n + m$ is the order of the higher order mode as indicated in Fig. 10.14.

10.1.2 Radius of Curvature

The radius of curvature ρ of the spherical surface of the cavity is related to the waist size of the transmitted beam by the following relationship:

$$w_0^2 = \frac{\lambda}{n\pi}\sqrt{2L_{geo}\left(\rho - 2L_{geo}\right)} \tag{10.2}$$

where $\lambda = 1064$ nm is the beam wavelength.

The radius of curvature has been chosen as a compromise to satisfy three criteria:

- to minimize the filtering of the higher order modes (HOMs) and the RF sidebands (SBs);

- to limit the scattered light noise;
- to keep the thermo-refractive noise a factor 10 below the sensitivity curve of AdV+ (see Sect. 2.2.3).

These criteria are briefly described in the following.

10.1.2.1 Higher Order Modes and Radio-Frequency Sidebands Filtering

The radius of curvature appears in Eq. 10.1 which describes the cavity transmission factor for both the carrier in its fundamental mode TEM00 and its higher-order modes TEMnm, but also for RF sidebands at different modulation frequencies, by setting different conditions.

The transmission of the carrier's higher-order mode TEMnm is obtained by setting the resonance condition $\frac{2\pi L_{opt} \nu}{c} = 0$ in Eq. 10.1 and fixing the order $N = n + m$.

The transmission of the RF sidebands is obtained by setting $\Delta\nu = 6$ or 56 MHz in Eq. 10.1.

10.1.2.2 Scattered Light Constraint

The scattered light produced by the OMC cavity surfaces is inversely proportional to the waist of the cavity squared w_0^2 (see Eq. 6.7) and therefore to the cavity radius of curvature ρ from Eq. 10.2.

10.1.2.3 Thermo-Refractive Noise Constraint

The thermo-refractive noise (δl_{tr}) is inversely proportional to the cavity waist squared and therefore to the radius of curvature, from Eq. 10.2. This noise is described by the relation [111]

$$\delta l_{tr} \sim \frac{2\beta\sqrt{L_{rt} K_B \mathcal{K} T}}{CD} \frac{2}{2\pi f \sqrt{\pi} w_0^2}. \tag{10.3}$$

The parameters used in the Eq. 10.3 are summarized in the Table 10.3.

Its coupling to the sensitivity is discussed in Sect. 10.1.3.

From these considerations, a radius of curvature of 1700 ± 8 mm was chosen, which corresponds to a beam of waist $321.0 \pm 0.4\,\mu$m. The tolerance on the radius of curvature takes also into account manufacturing error bars.

Table 10.3 Parameters used in Eq. 10.3

Parameter	Symbol	Value
Temperature dependence of the refractive index	β	$-1.0 \times 10^{-5}\mathrm{K}^{-1}$
Substrate density	D	$2200\ \mathrm{kg/m^3}$
Geometrical round-trip length	$L_{rt} = 4L_{geo}$	248 mm
Thermal conductivity	$\mathcal{K}$	$1.38\ \mathrm{Wm^{-1}K^{-1}}$
Boltzmann constant	K_B	$1.38 \times\ 10^{-23}\mathrm{m^2\ kg^{-1}\ s^{-2}K^{-1}}$

Table 10.4 OMC surfaces coating requirement. The surfaces are defined in Fig. 10.2: 1 is the input surface, 4 is the output surface and 3 is the curved surface. Surfaces 1 and 4 are defined in order to have the required finesse. Surfaces 2 and 3 are coated as best as possible to minimize losses

Surface	Requirement	Measurement at LMA
Surface 1, 4	$R = 0.9970 \pm 0.0001$	$R = 0.9970 \pm 0.0001$
Surface 2, 3	$R = 1$	$T = 1.6$ ppm

10.1.3 Finesse

The finesse of a bow-tie cavity, whose surfaces have reflectivities ~ 1, is given by

$$\mathcal{F} = \frac{\pi\sqrt{r_1 r_2 r_3 r_4}}{1 - r_1 r_2 r_3 r_4} \tag{10.4}$$

where r_1, r_2, r_3, r_4 are the reflectivities of the mirrors of the cavity.

The values of these reflectivities for the new high finesse cavity are reported in Table 10.4 [112] and from Eq. 10.4, the finesse of the cavity is 1032 ± 34.

A higher finesse with respect to O3 was needed to be able to filter the 6 MHz sideband (following Eq. 10.1), that was not needed before. However, the value of the finesse cannot be too high in order not to increase too much the internal losses and the cavity length noise. Therefore it has been chosen as a compromise between these different constraints:

- keep the internal losses below 2%;
- keep the length noise well below the AdV+ sensitivity curve;
- filter the RF sidebands.

10.1.3.1 Internal Losses Constraints

The internal losses of the OMC cavity are amplified by the number of round-trips done by the beam inside the cavity. This number is proportional to the finesse. Considering the internal losses contributions it has been estimated that the finesse cannot be higher than 1000 to be within 2% of internal losses, as it will be detailed in Sect. 10.3.4.

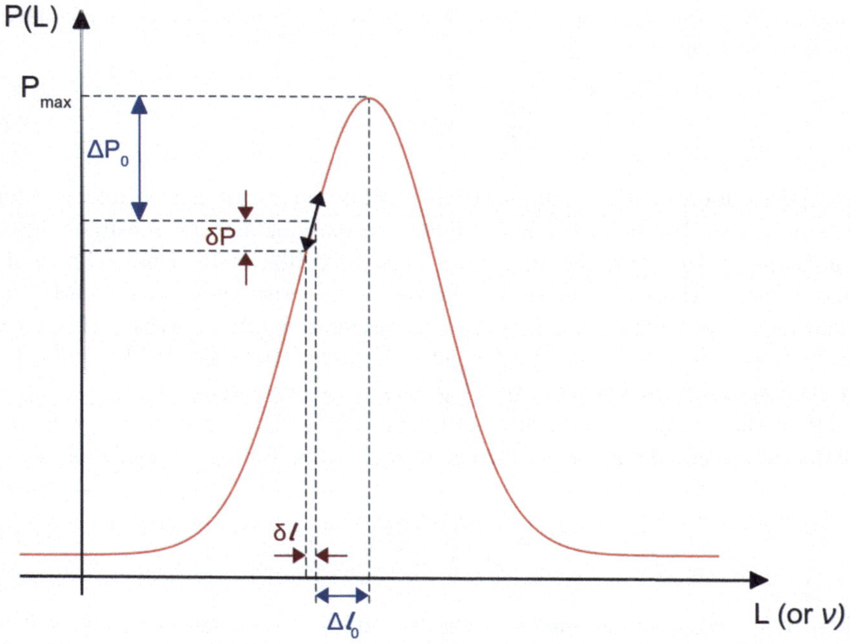

Fig. 10.3 Power transmitted by the OMC cavity. The static cavity length variation δl_0 is given by the lock precision of the cavity and it is seen as a power reduction δP_0 in the transmitted power. The dynamic cavity fluctuation $\delta l(f)$ is the cavity length noise and it is seen as power fluctuation δP in the transmitted power. The dependency of the power can be also be expressed in terms of frequency ν as shown in Sect. 2.1.2

10.1.3.2 Cavity Length Noise Constraints

The coupling of the OMC length noise is proportional to the square of the finesse as will be explained here. The length noise of the OMC results in a variation of the OMC transmitted power that pollutes the gravitational wave signal, especially if the cavity is not exactly at the resonance, as shown in Fig. 10.3. The coupling of this noise to the GW signal depends on the precision of the OMC cavity locking. The deviation in length of the cavity with respect to the strict resonance condition (lock precision) is indicated with Δl_0 (largely exaggerated for illustration in Fig. 10.3) and corresponds to a power difference ΔP_0 with respect to the maximum transmitted power P_{max}, as indicated in Fig. 10.3. The variation $\delta l(f)$ around the working point given by the lock precision is the cavity length noise and corresponds to an additional power variation $\delta P(f)$.

The power transmitted by the OMC cavity is given, accordingly to Eq. 2.17, by

$$P_T = P_{max} - \Delta l_0 - \delta P = \frac{P_{max}}{1 + \left(\frac{2\mathcal{F}}{\pi}\right)^2 \sin^2\left(\frac{2\pi}{\lambda}\left(\Delta l_0 + \delta l\right)\right)}. \tag{10.5}$$

From Eq. 10.5, one gets for $\Delta l_0 \quad << \lambda$ a cavity length noise, in terms of relative power fluctuation, of

$$\frac{\delta P}{P} = 32\mathcal{F}^2 \frac{\Delta l_0}{\lambda} \frac{\delta l}{\lambda}. \tag{10.6}$$

The specification of the maximum cavity length noise that can be tolerated has been chosen so that the projection of this noise onto the detector sensitivity curve is a factor of 10 below the shot noise curve over the entire bandwidth of the interferometer. The lock precision is the rms of the error signal of the OMC and it constant. The length noise includes the thermo-refractive noise δ_{tr} defined in Eq. 10.3. This noise is frequency dependent between 10 Hz and 10 kHz and it has been measured during O3 not to spoil the sensitivity of the detector.

The replacement of two cavities of finesse 123 with a single cavity of finesse 1000, is increasing the impact of the length noise, from Eq. 10.6, by a factor Δ

$$\Delta = \frac{1}{\sqrt{2}} \left(\frac{1000}{123} \right)^2 = 45 \tag{10.7}$$

where $1/\sqrt{2}$ takes into account the fact that during O3 we had two cavities. This worsening factor can be, however, compensated by the possibility of improving the lock precision Δl_0 of the cavity. Indeed, the OMC lock precision has been measured at LAPP [110] to be

$$\Delta l_0 \sim 6 \times 10^{-13}\text{m}. \tag{10.8}$$

During the run O3, it has been measured that the lock precision of the two cavities was $\Delta l_0 \ \sim 4 \cdot 10^{-11}$ m [113]. This means that the lock precision can be improved by a factor 100. As shown in Fig. 10.3, the lock precision can be expressed also as a frequency variation $\Delta \nu_0$. This means that by improving the stabilization of the Virgo laser frequency (by acting on the control filter), the lock precision can be improved until we reach the value measured at LAPP (see Eq. 10.8). Assuming this improvement, for O4 the OMC length noise will be a factor 5 below the design sensitivity at 30 Hz and a factor 10 at 100 Hz, that is within the requirements.

10.1.3.3 Radio-Frequency Sidebands Filtering

The cavity finesse was chosen in such a way as to appropriately filter the incoming sidebands so as not to increase the shot noise level. During O3 only the 56 MHz sidebands had to be filtered out, while in O4 the 6 MHz sidebands, which are necessary to control the Signal Recycling cavity (see Sect. 2.1.4), also had to be filtered out, requesting a higher cavity finesse. In Fig. 10.14, it can be seen from the

Table 10.5 Sideband parameters and simulation results [114]. The modulation index m indicated in the table refers to the maximum value that can be produced by the EOM. The values of the output values are given by the simulations. The OMC transmission is given for a finesse of 1000

Sideband frequency	Modulation index	ITF output power	OMC transmission	B1 power
Carrier	–	5 mW	1	5 mW
6.27 MHz	0.25	10 mW	4.4×10^{-3}	44 μm
56.44 MHz	0.25	599 mW	5.5×10^{-5}	33 μm

cavity transmitted modes how close is the 6 MHz sideband to the carrier, stressing the need of properly filter not to introduce noise.

The requirement we want to meet is that the shot noise ($\propto \sqrt{P}$) due to the amplitude fluctuation of the sideband at 6 MHz (h^{SB}_{shot}) is 5 times lower than the shot noise of the carrier (h^{car}_{shot}).

The attenuation factor T_{SB} is given by Eq. 10.1, by setting $N = 0$ and $\Delta nu =$ 6 MHz. The sideband power transmitted by the OMC can be computed from

$$P_{T_{SB}} = P_{in} \cdot \frac{m^2}{2} \cdot T_{SB} \tag{10.9}$$

where P_{in} is the input power (40 W during O4) and m is the modulation index, defined in Sect. 2.4.2. The parameters relative to O4 are reported in Table 10.5.

Using the values in Table 10.5, the ratio between the carrier and the sideband shot noises, without considering the squeezing injection, is

$$\frac{h^{car}_{shot}}{h^{SB}_{shot}} \propto \frac{\sqrt{P_{T_{car}}}}{\sqrt{P_{T_{SB}}}} = \frac{\sqrt{5\text{mW}}}{\sqrt{44\mu\text{W}}} \sim 10. \tag{10.10}$$

The shot noise of the carrier is however lowered by the presence of squeezing. In AdV+, we expect up to 4.5 dB of shot noise reduction [74], leading to a reduction factor of $10^{4.5/20} = 1.7$

$$\frac{h^{car}_{shot}}{h^{SB}_{shot}} \propto \frac{\sqrt{P_{T_{car}}}}{\sqrt{P_{T_{SB}}}} = \frac{\sqrt{5\text{mW}} \cdot 1.7}{\sqrt{44\mu\text{W}}} \sim 6. \tag{10.11}$$

In conclusion, the shot noise introduced by the power fluctuations of the 6 MHz sidebands transmitted by the new OMC is 6 times lower the AdV+ shot noise which is in agreement with the requirements. Moreover, the power of the 56 MHz sidebands after the OMC, according to the values in Table 10.5, is of the same order of magnitude of the 6 MHz sidebands which is therefore also low enough.

10.2 Output Mode Cleaner Cavity Mechanical Support and Control

The cavity lies on an aluminum stand, as during O3. The cavity length needs to be controlled with a feedback control loop (see Sect. 2.4.1). The control loop correction is sent both to Peltier cells, which act on the cavity temperature, and to a piezoelectric actuator, which imparts mechanical compression to the cavity. These actuators are shown in Fig. 10.4: the two Peltier cells are located below the copper plate on which the cavity is placed and the piezo-electric actuator ($5 \times 5\text{mm}^2$) is located above the cavity, held by a screw.

The temperature variation changes the optical path length by changing the refractive index $n = f(T)$ of the cavity material. The response time of the Peltier cells is about 10 seconds. Therefore this kind of control is slow but very stable.

The piezo-electric actuator changes the geometrical length of the cavity by compressing it from the top. It is used as a complementary actuator for the control of the cavity length, as its response in time is much faster than that of the Peltiers. The screw holding the piezo is screwed into an invar bridge with low thermal expansion.

An aluminum cover is placed around the cavity, in order to perform a heat shield, to protect it from handling around the bench and to prevent scattered light coming from optics around to enter it.

Fig. 10.4 OMC cavity mounted on its support. On the bottom there are two Peltier cells used for the temperature control and on the top there is a piezo-electric screw (PZT) used for the mechanical control. Credit P. Spinicelli

10.3 Output Mode Cleaner Cavity Characterization at LAPP

Before installing the new OMC on site, we tested and characterized two high finesse cavities in the optical laboratory at LAPP, in order to chose the one with the best performance. In the following, we describe the general procedure for characterizing these cavities and give the results concerning the cavity we have chosen to install on the site.

10.3.1 Output Mode Cleaner Cavity Test Optical Set-Up

The set-up used to test the OMC cavities is shown in Fig. 10.5 and uses two cavities: the high-finesse to be tested and a low-finesse.

The low finesse OMC cavity is a spare OMC cavity, like the ones used during the observing run O3, with finesse of 123. It is used to stabilize the laser frequency which is needed for the characterization of the high finesse cavity. The laser used for the experiment is a solid state Nd-YAG, with 1064 nm of wavelength. The maximum power emitted by the laser is 1 W and it is easily tuned by changing the current

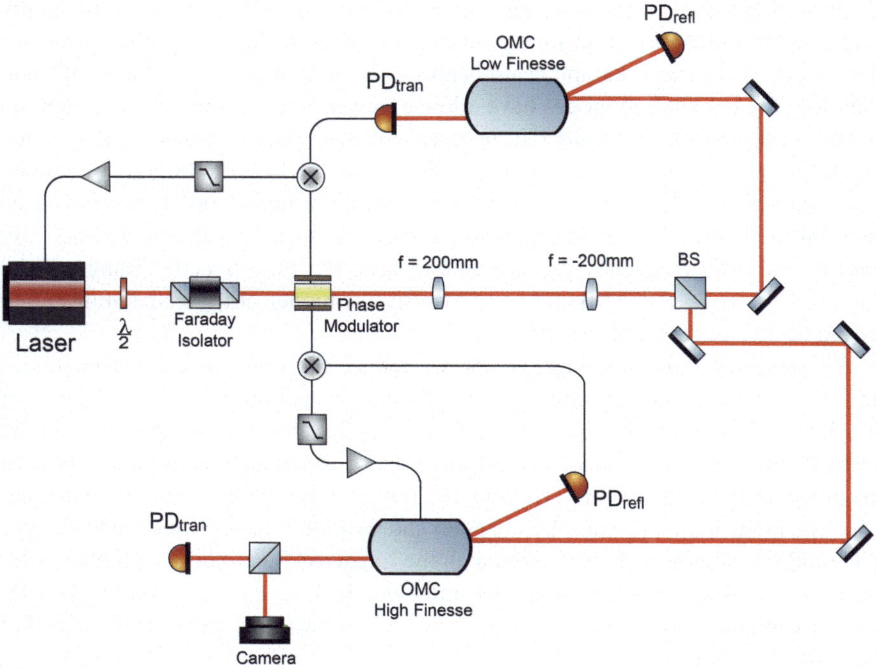

Fig. 10.5 Simplified scheme of the set-up on the bench in the optics laboratory at LAPP. The two steering mirrors in front of the high finesse cavity are used to align the OMC

directly from the laser controller or by tuning the half wave plate after the laser and before the Faraday Isolator. The Faraday Isolator is needed to have a beam s-polarized going to the OMC cavity (since it has two polarizers, one at the input and one at the output) and to avoid back-reflected light to go back to the laser, avoiding to damage it. Then there is a telescope composed by two lenses (of focal lengths $f_1 = 200\,\text{mm}$ and $f_2 = -200\,\text{mm}$) used to match the laser beam to the OMC in order to have the beam waist $w_0 = 320\,\mu\text{m}$ at the cavity.

10.3.2 Alignment of the Output Mode Cleaner Cavity

The first thing we did after installing the monolithic cavity was to align it (see Appendix A). The alignment is performed acting on the two steering mirrors in front of the cavity. As figure of merit for the alignment, the shape of the transmitted beam from the cavity is observed. We vary the laser frequency in order to see the shape of the beam transmitted with a IR laser detection card. The shape of the transmitted beam looks like that of an ellipse. In fact, the misaligned beam has higher order modes which, depending on their orientation, tell us whether our beam is vertically or horizontally misaligned within the cavity. So if the ellipse we see in transmission is oriented more vertically than horizontally, we act on the vertical degrees of freedom of the steering mirrors to make the ellipse more horizontally oriented. We iterate this procedure until the size of the major axis of the ellipse has decreased. This means that the beam begins to be well aligned with the OMC, and therefore the transmitted modes have a lower power and are difficult to be seen on the IR card. Therefore, in order to proceed with the fine alignment of the cavity, we put in transmission a beam splitter which sends the beam to both a photodiode and a camera, as shown in Fig. 10.5. In this way the higher order modes can be seen on the camera and their respective power monitored by the photodiode. By varying the laser frequency, we are able to scan the Free Spectral Range of the cavity ($FSR_{OMC} = 834\,\text{MHz}$) and to identify the higher order modes transmitted by the cavity, as shown in Fig. 10.6.

By using the same steering mirrors, we reduce the order of the higher modes and their power. Again, to reduce a vertical mode we act on the vertical degrees of freedom and the same for the horizontal ones. Once almost all the power is on the TEM00, the cavity is aligned. The misalignment is computed as the ratio between the power of the first Hermite-Gaussian (HG) high order mode and the fundamental one. The mode matching was already good and we didn't need to tune the telescope. The mode mismatch is characterized with the Laguerre-Gaussian (LG) higher order modes. The OMC alignment is further fine tuned locking the cavity on the TEM00 and maximizing the transmitted power. The lock of the OMC cavity is described in the following section.

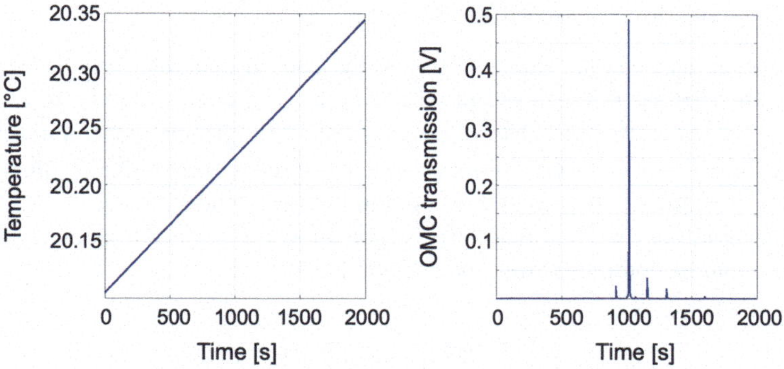

Fig. 10.6 Alignment of the OMC cavity. On the left, the modulation of the laser frequency performed by changing the temperature. On the right, the cavity transmission detected by the photodiode: the fundamental mode TEM00 in the middle and higher order mode around

10.3.3 Lock of the Output Mode Cleaner Cavity

Once the cavity is aligned, its length must be controlled so that the TEM00 mode resonates, i.e. the cavity must be locked, see Sect. 2.4. The frequency of the laser we use for our experiment fluctuates a lot due to thermal variations, mechanical vibrations and changes in the length of the laser cavity for example. So before locking the high finesse OMC cavity, we took care of stabilizing the laser frequency by using an additional auxiliary cavity, i.e. the low finesse OMC. The error signal of this loop is derived with the Pound-Drever-Hall (PDH) technique, described in Sect. 2.4.2. The beam is modulated at 3231.3 Hz by acting on the laser frequency instead of using the electro-optical modulator. The actuation of the frequency stabilization loop was sent to the laser crystal). The scheme of the laser frequency stabilization loop, using the OMC low finesse as a reference, is shown on the top of Fig. 10.5. Once the laser frequency is stabilized, we can control the length of the OMC high finesse cavity, using the same PHD technique. We modulate the OMC cavity length with the piezo-electric actuator (see Sect. 10.2) at 1959 Hz. In this case the feedback is sent directly to the OMC actuators (Peltiers and piezo), in order to act on its length. The cavity has been locked acting first only on the Peltier cells and then also on the PZT, to have a lock in full-bandwidth. The scheme of the high finesse OMC cavity length control loop is shown on the bottom of Fig. 10.5.

10.3.4 Internal Losses

The parameters of the OMC (geometry, materials, etc.) have been optimized to satisfy the constraints on the HOMs and sidebands filtering but also to minimize losses and noises.

OMC internal losses have different origins:

1. absorption losses in the substrate due to the chosen material. The absorption of Suprasil 3001 is (0.3 ± 0.2) ppm/cm, which for a cavity round trip length of $4 \cdot L_{geo} = 24.8$ cm gives absorption losses of (7.5 ± 5) ppm per round trip.
2. absorption losses in the coatings has been measured to be less than 2 ppm ((1 ± 1) ppm) per coating [110], so (4 ± 4) ppm in total per round trip.
3. scattering losses due to mirror defects (polishing and surface treatment) for each surface can be computed from the relation [115]

$$L_{rough} = \left(\frac{4\pi\sigma_{RMS}}{\lambda}\right)^2 \sim (0.7 \pm 0.4)\text{ppm} \tag{10.12}$$

   where $\sigma_{RMS} = (0.07 \pm 0.01)$ nm RMS [116], see Sect. 10.1.1. Considering the 4 surfaces of the cavity, the scattering losses are (2.8 ± 1.6) ppm per round-trip. This is, however, a lower value than what was measured at LMA [117] of 3.5 ± 0.5 ppm per surface which corresponds to 14 ± 2 ppm for the four faces. This value is higher because it takes into account point defects and other imperfections that are neglected by the value obtained from Eq. 10.12. In estimating total losses we will consider scattering losses equal to 14 ± 2 ppm per round trip as it is more realistic.
4. the Rayleigh scattering has been measured to be (0.70 ± 0.02) ppm/cm [97] for fused silica therefore, for a round trip, is (17.0 ± 0.5) ppm.
5. the HR mirror residual transmission accounts for only (2.5 ± 0.4) ppm of losses.

The overall value of the round-trip losses is given by the sum of all the contributions and it is given by (45 ± 12) ppm, as summarized in Table 10.6. The round-trip losses must be multiplied by the number of round trips defined as

$$N = \frac{\mathcal{F}}{\pi} \sim 318 \qquad \text{for} \quad F = 1000. \tag{10.13}$$

The total expected losses are therefore $(1.4 \pm 0.4)\%$.

In addition, there are astigmatism losses, estimated to be (50 ± 5) ppm for the previous low finesse cavities [118]. These losses are intrinsic to the geometry of cavity, so they are assumed to be the same. Its contribution is negligible compared to the other losses.

Table 10.6 Internal loss contributions per round-trip

Losses sources	Value per round-trip
Suprasil absorption	(7.5 ± 5) ppm
Coating absorption	(4 ± 4) ppm
Surface scattering	(14 ± 2) ppm
Rayleigh scattering	(17.0 ± 0.5) ppm
HR transmission	(2.5 ± 0.4) ppm
Total	(45 ± 12) ppm

The power losses L are defined as $L = 1-R-T$, where R and T are the reflected and transmitted powers (P_R and P_T) normalized by the cavity input power P_I. The power reflected, transmitted and at the input of the cavity have been measured using an integrated sphere to have a more precise measurement compared to a powermeter. The sphere has been calibrated using the powermeter and we verified that its response was linear in the range of powers we are measuring. To estimate the OMC losses, we measured the powers P_I, P_R and P_T with the integrating sphere. The sphere requires the beam to enter at a negligible angle of incidence in order to make an accurate measurement of its power. For this reason, we made sure that the beam propagated along the optical bench for a length of about 60 cm before entering the sphere, so that it could be aligned with the detection axis. For each power value, we performed several power measurements at the same position, re-positioning the sphere on the beam and taking the standard deviation of the measurements as error.

The measured value of internal losses is $L = (2.89 \pm 0.13)\%$. Since the expected internal losses of the cavity are expected to be $\leq 2\%$, we cleaned the cavity with a cleaning solution composed by a polymer called *First Contact*. We repeated the measurement twice and we obtained $L = (2.12 \pm 0.13)\%$ and $L = (1.98 \pm 0.06)\%$ meaning that the cleaning reduced the losses. The average of the two results is $(2.05 \pm 0.10)\%$, which is slightly higher than the theoretical estimation. The small discrepancy between the two values can be due to surface defects and residual dust contamination.

10.3.5 Finesse

The finesse has been chosen for the new output mode cleaner to be 1000 in order to allow a good filtering of the 6 MHz sidebands. To measure the finesse value of the cavity we scanned it over its entire spectral range and we acquired the power in transmission of the cavity, see Eq. 10.1. The FSR scan of the OMC can be performed acting on two different observables of Eq. 10.1:

- the laser frequency ν;
- the temperature T of the OMC cavity (as explained in Sect. 10.2)

here we use the temperature.

In Fig. 10.7, the temperature scan of the cavity is shown: on the left the temperature variation read by a thermometer in contact with the cavity and the right the OMC transmission, showing a free spectral range (see Sect. 2.1.2). We left the cavity scanning for several days to have more statistics and extract a precise value of the finesse.

From the scans one can derive the value of the FSR and linewidth (FWHM) of the OMC cavity, see Eqs. 2.23 and 2.24 in Sect. 2.1.2 and therefore one can compute the value of the finesse.

We analyzed the scans of the OMC. It was not possible to analyze more that 24 hours of data at the time so we took three different datasets and we performed the

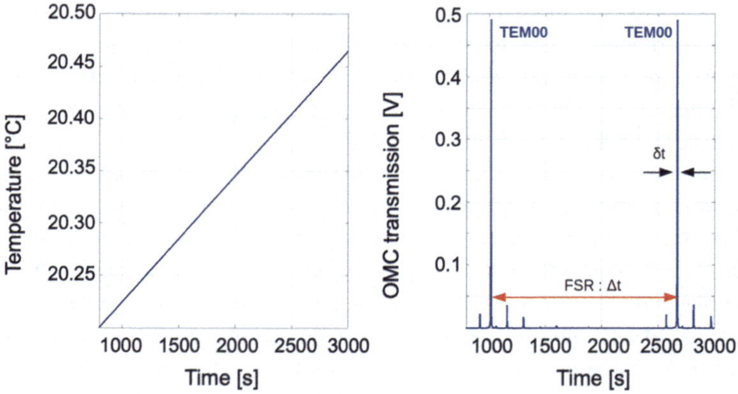

Fig. 10.7 OMC cavity scan, acting on the laser frequency (actuating on the laser crystal temperature). On the left the modulation of the laser frequency performed by changing the temperature. On the right the power transmitted by the OMC: the Free Spectral Range (FSR: ΔT) is the distance between two consecutive resonances, in this case of the fundamental mode TEM00, δt is the width of the peak in correspondence of the half of the TEM00 (FWHM)

Fig. 10.8 Histogram for the finesse of the OMC, calculated from of the three datasets

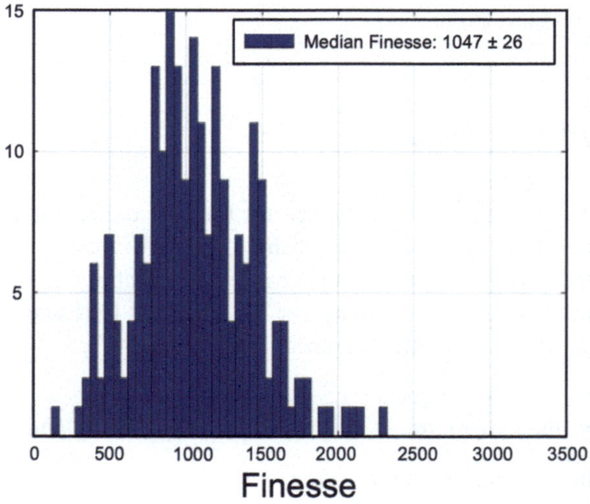

average. Figure 10.8 shows one of the three datasets. The medians of the histograms give the following finesse: (1047 ± 26) mm, (1040 ± 27) mm and (1080 ± 26) mm. The average of these measurements is (1055 ± 15) mm, considering the error to be statistical, which is consistent with the specifications.

10.3.6 Radius of Curvature

The radius of curvature has been measured with two different methods. The first one using the waist of the beam transmitted by the OMC, according to the relation of Eq. 10.2, and the second one using the frequency scan of the cavity.

10.3.6.1 Measurements from the Waist of the Transmitted Beam

From Eq. 10.2 we can derive the radius of curvature of the cavity as

$$\rho = \frac{n^2 \pi^2 w_0^4}{2\lambda^2 L_{geo}} + 2L_{geo}. \tag{10.14}$$

From this formula we can derive ρ from the measurement of the waist of the beam transmitted by the OMC. To have a more stable lock of the OMC, we locked the laser on the high finesse mode cleaner cavity. We measured the beam size at different positions along the beam propagation. We placed the camera on the beam and then we averaged 60 seconds of data for several positions. For every acquired image of 60 seconds, we fit the image profile (with a Gaussian) and derive the beam size. After having determined the value of the beam size for every position of the camera, we used the same procedure explained in Sect. 5.1.1. The beam transmitted by the OMC cavity, as shown in Fig. 10.9, is described by the following parameters:

- $w_{0x} = (330 \pm 15)\ \mu m$ at $z_{0x} = (38 \pm 15)$ mm;
- $w_{0y} = (340 \pm 9)\ \mu m$ at $z_{0y} = (29 \pm 9)$ mm.

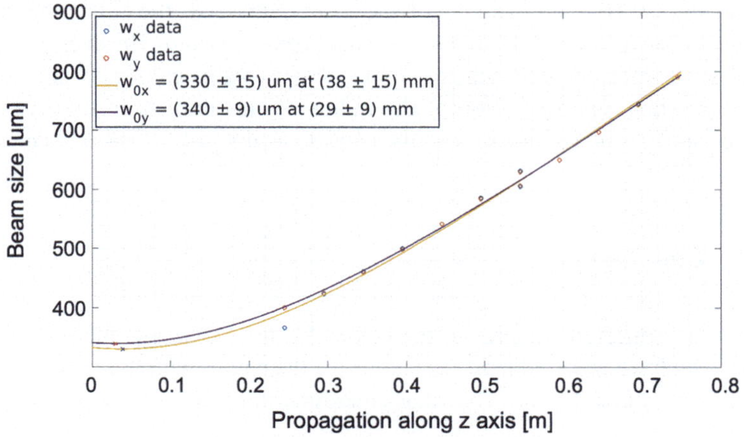

Fig. 10.9 Fit to derive the waist size of the beam transmitted by the OMC. The blue points are the w_x data and the yellow line is the curve fitting those points. The red points are the w_y data and the purple line is the curve fitting those points

Using the Eq. 10.2, we obtain:

- $\rho = (2079 \pm 300)$ mm (using w0x);
- $\rho = (2225 \pm 200)$ mm (using w0y).

Since there is a dependence on the fourth power of the waist on that formula, the uncertainty on ρ using this method is quite large.

10.3.6.2 Measurements from the Frequency Scan of the Cavity

Using the same scans of the OMC cavity performed to measure the finesse, we could estimate also the radius of curvature of the cavity. According to Eq. 10.1, a mode of order N is at resonance when

$$\frac{2\pi \, \Delta\nu L_{opt}}{c} - N \arccos\left(\sqrt{1 - \frac{2L_{geo}}{\rho}}\right) = 0 \tag{10.15}$$

where $\Delta\nu$ is the distance in frequency between the fundamental mode and an order N mode, for example TEM00 and TEM01, $\Delta\nu = \nu_{00} - \nu_{01}$. From Eq. 10.15 one gets

$$\rho = \frac{2L_{geo}}{\sin^2\left(\frac{2\pi \, L_{opt}}{c} \Delta\nu\right)} \tag{10.16}$$

By acquiring a long period of scan data, we measured $\Delta\nu$. We divided the data in several datasets and then we compute the average of the median value of each of the three datasets. The histograms of one of the three datasets is shown in Fig. 10.10. The medians of the histograms give the following radii of curvature: (1686 ± 3) mm, (1686 ± 4) mm and (1684 ± 4) mm. The average of these measurements is (1686 ± 2) mm, where the error is statistical. This value is much more precise than the estimation of the radius of curvature from the transmitted beam size and it is in agreement with the expected value $\rho = (1700 \pm 8)$ mm.

10.3.7 Conclusions

We tested two different cavities in the laboratory to choose the one with the best substrate. We have reported here the values for the cavity with the best results that was then installed on the site. The values measured for this cavity were:

- Losses = $(2.05 \pm 0.10)\%$ which is slightly higher than the theoretically calculated $(1.4 \pm 0.4)\%$ but within the specifications;

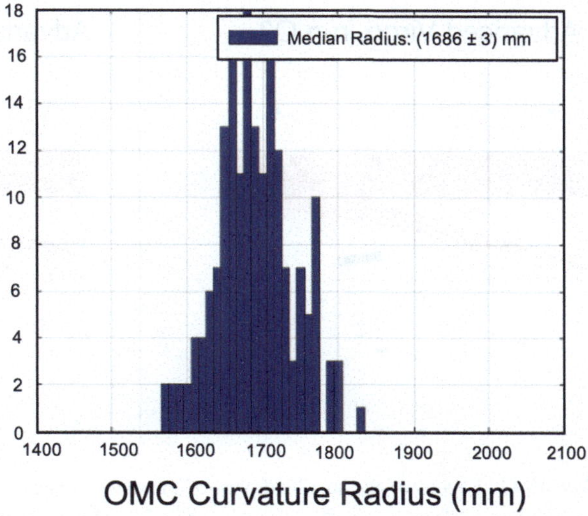

Fig. 10.10 Histogram for the radius of curvature of the OMC, calculated from one of the three datasets that we averaged

- Finesse = 1055 ± 15 in agreement with the finesse expected from the surface reflectivities 1032 ± 34 and the specifications;
- Radius of curvature = (1689.3 ± 2.3) mm (from scanning the cavity) in agreement with the design value (1700 ± 8) mm.

The parameters obtained are in agreement with what we expected from the specifications. Therefore, we approved the cavity characterized at LAPP and it was decided to install it on Virgo.

10.4 Output Mode Cleaner Cavity Installation and Pre-commissioning

In November 2020 we installed the new cavity on the SDB1 detection bench, replacing the two low finesse cavities used during the previous observing run. The cavity was installed on one of the two supports already on the bench while the support of the second cavity was removed. Several modifications were made to the optical set-up on the bench to allow the new cavity to be integrated. After the hardware intervention, we proceeded with the alignment and matching of the cavity with the interferometer beam. Misalignment and mode mismatch count as detection losses for the squeezing. In this section, we report the results obtained concerning these losses during the commissioning for O4.

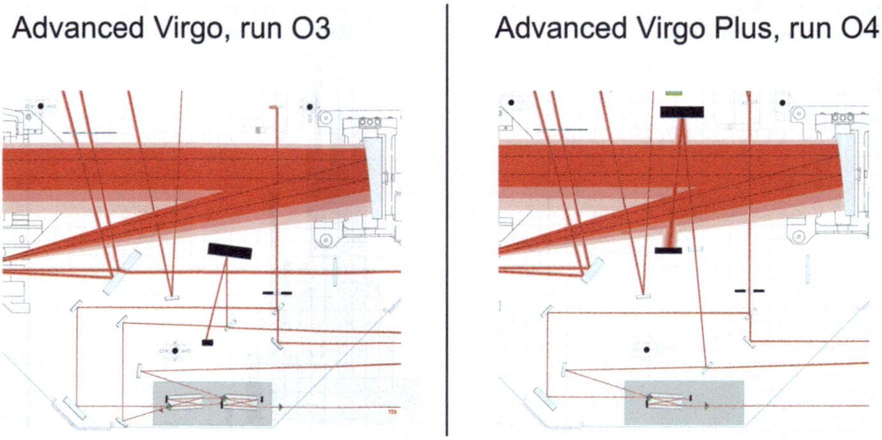

Fig. 10.11 Layout of the modified area of the detection optical bench SDB1. On the left, the configuration during the observing run O3, with two OMC cavities. On the right, the configuration for the O4 observing run, with only one OMC cavity

10.4.1 Optical Configuration Change

In Fig. 10.11 the optical set-up of the bench SDB1 during the previous observing run O3, on the left, and the optical configuration that will be used during the observing run O4. The main difference is that instead of two cavities there is only one cavity, meaning that we don't have two reflected beams coming from the cavities to be sent to the detectors. Since the power is high on the reflected beam, only a pick-off of the reflected beam is detected, while most of the power is going on a beam dump.

Learning from past experience, it was also decided to install two motorized wave plates to fine-tune the polarization of the incoming beam in the OMC, as shown in Fig. 10.12 (left). As the telescope mirror is very large and therefore heavy, moving it along the beam axis, on its translation stage, leads to an unbalance of the suspended bench. In order to balance it remotely, we installed a motorized counterweight under the bench, as shown in Fig. 10.12 (right).

10.4.2 Output Mode Cleaner Cavity Alignment on SDB1

The OMC cavity has been aligned following the same procedure used at LAPP as described in Sect. 10.3.2. We started with the SDB1 bench tower open, i.e. being able to see the transmitted beam with an infrared card. We worked with the bench suspended and controlled, acting on the picomotors on the two mirrors in front of the OMC to move the beam horizontally and vertically. Once the shape of the output beam was a sufficiently Gaussian, we began to look at sensors positioned

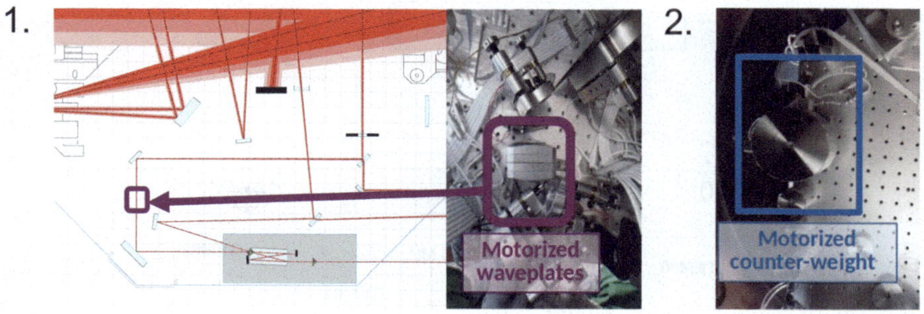

Fig. 10.12 On the left the motorized wave-plates installed before the OMC cavity on SDB1 to tune the polarization matching. On the right the motorized counter-weight installed below SDB1 bench for the fine tuning of the OMC telescope, i.e. to balance the movement of the big meniscus lens along the beam axis placed on SDB1

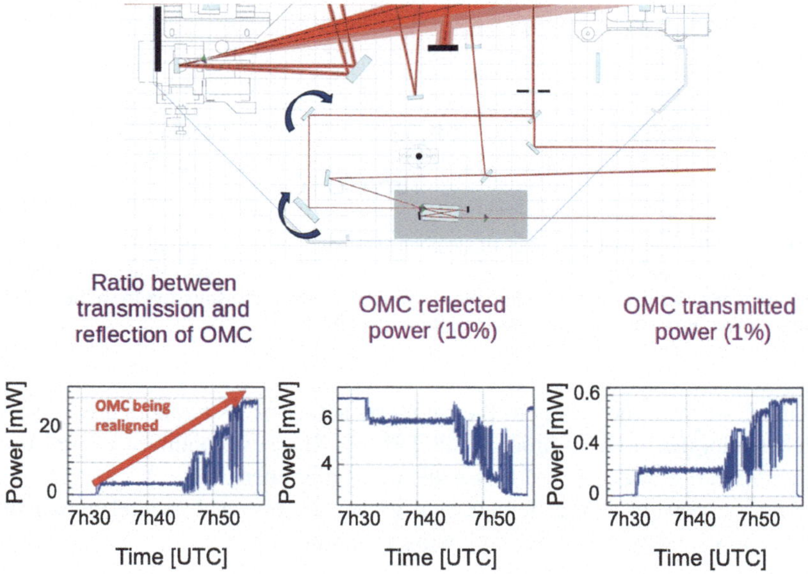

Fig. 10.13 Alignment of the OMC cavity acting on the positions of the last two mirrors before the cavity, using the picomotors (indicated by blue arrows on the upper design). On the bottom, the alignment improvement seen by the powers on the photodiodes: the plot on the left represents the ratio between the transmitted and reflected cavity powers, the middle plot shows the decrease in reflected power, and the right plot displays the transmitted power

on SDB2, the bench that follows SDB1 and on which the cameras, photodiodes are located. The beam transmitted by the OMC, is the one containing the gravitational waves signal, called B1. We maximize the ratio between the transmitted and the reflected powers of the OMC cavity by rotating the mirrors just before the OMC (with picomotors), as shown in Fig. 10.13. By doing this, we minimize also the reflected power and maximized the transmitted power. To calculate the alignment of

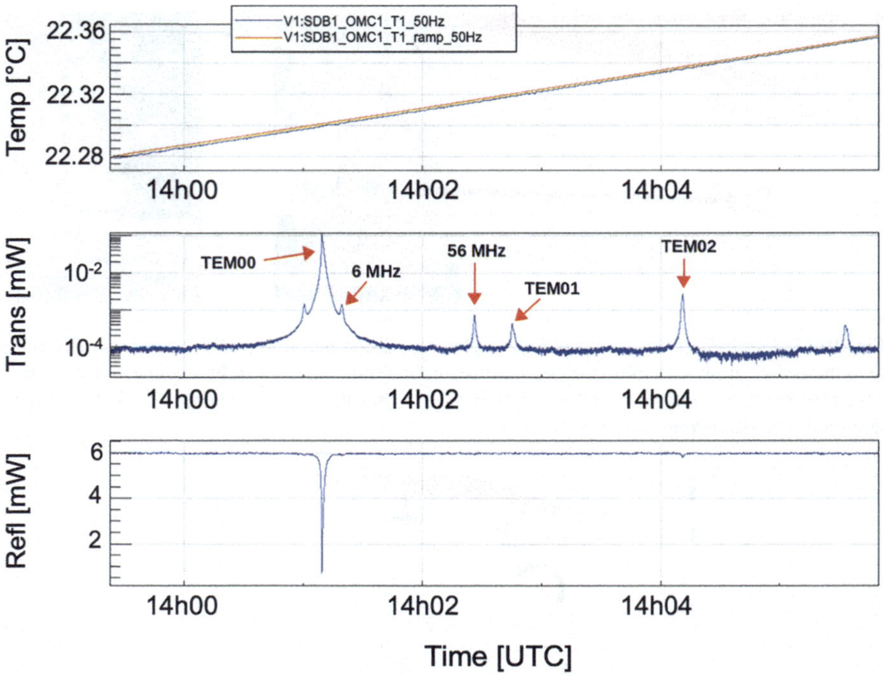

Fig. 10.14 OMC temperature scan in single bounce configuration. We checked the alignment and mode matching by looking at the power on the first and second higher order modes. The upper plot shown the temperature scan, the middle plot shows the signal in transmission from the OMC and the bottom one shows the signal in reflection from the OMC

the OMC, we scan the length of the cavity by acting on its temperature, to see the ratio between the mode of order 1 and TEM00. The best condition we reached in this configuration was 0.4% of losses due to misalignment.

From the powers of the fundamental mode, the first and second higher order modes, shown in Fig. 10.14, the following values are derived

- Misalignment losses $\sim 0.4\%$;
- Mode mismatch losses $\sim 2.5\%$.

We then worked on reducing the mode mismatch of the interferometer beam with the OMC cavity.

10.4.3 Output Mode Cleaner Cavity Mode Matching on SDB1

The ITF incoming beam must be mode matched with the OMC cavity, to avoid to introduce mismatch losses. This is done acting on the telescope elements on SDB1. The last interferometer mirror that the beam encounters is the Signal Recycling

(SR) and then it arrives at the SDB1 telescope composed of a meniscus lens (of diameter 150 mm) and two parabolic mirrors (MMT_M1, of diameter 150 mm and MMT_M2, of diameter 30 mm). To improve the matching we tried two different strategies: acting on the distance between SR and the meniscus lens or acting on the position of the parabolic mirrors.

10.4.3.1 First Method: Distance Between Signal Recycling Mirror and Meniscus Lens

At first, we improved the mode matching by increasing the distance between SR and SDB1 meniscus lens, as shown in Fig. 10.15.

The motorized counterweight has been used to compensate for the meniscus lens movement keeping the suspended bench balanced and controlled. We optimized the matching by maximizing the ratio between transmitted and reflected OMC powers and minimizing higher order mode amplitude. The best situation with this method has been obtained by displacing the meniscus lens by +4.5 mm. Since we could not further displace the lens, we moved the bench in the beam direction (z-axis) by +5.5 mm and the SR mirror by −3.0 mm. The total variation on SR-meniscus lens distance was $4.5 + 5.5 + 3 = 13$ mm. At this point the misalignment losses were 1.7% and the mode mismatch was 0.64%, mostly due to astigmatism. The signals relative to the matching improvement are shown in Fig. 10.16. It is not possible to leave these positions of SR and SDB1 during the science run as SR position should be set by the tuning of Signal Recycling Cavity (SRC) cavity length, and a large mis-centering of SDB1 may induce extra control noise. It was therefore decided to act on the other components of the telescope.

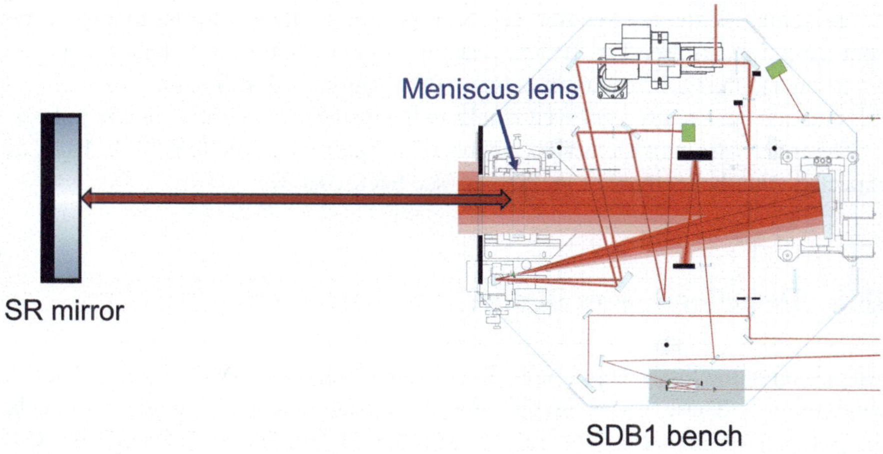

Fig. 10.15 First method to improve the mode matching of the beam with the OMC: to change the distance between the Signal Recycling mirror (SR) and the meniscus lens. Not on scale

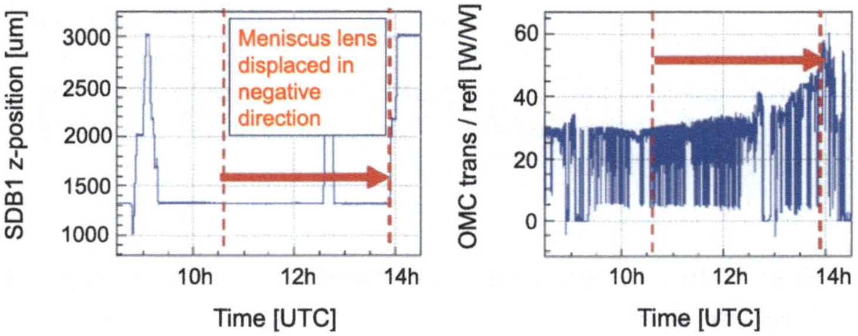

Fig. 10.16 Mode matching improvement by adjusting the distance between SR and the telescope. The left plot shows the z-position of the bench SDB1. The right plot shows the increase of the ratio between the transmitted and reflected powers, meaning that the matching is improving

10.4.3.2 Second Method: Displacement of the Parabolic Mirror

We restored the nominal positions of the bench SDB1 along z-axis (−5.5 mm) and SR mirror (+3 mm). The second method to improve the mode matching consists in longitudinally displacing the parabolic mirror MMT_M1 (and angularly MMT_M2 to correct for the tilt of the beam induced by the MMT_M1 displacement), as shown in Fig. 10.17. The optimal situation obtain with this method has been a misalignment of 0.5% and a mode mismatch of 2%, see Fig. 10.17.

We concluded that it was good enough as a pre-mode matching.

10.4.4 Polarization Tuning and Sideband Filtering

A fine tuning of the polarization has been performed thanks to the tuning of the polarization plates (see Fig. 10.12), reaching a polarization mismatching between the beam and the OMC cavity of only ∼0.08%. During O3, this value was ∼2%, so the situation has greatly improved thanks to the installation of the motorized plates.

The transmission of the 6 MHz sideband has been measured to be $3.7 \cdot 10^{-3}$, as expected with the increase of the finesse (see Sect. 10.1.3).

10.5 Detection Losses Budget

We measured the losses between the interferometer and the OMC, in single bounce configuration. These losses account for the misalignment, the mode mismatch, the polarization detuning. The values are reported in Table 10.7 for O3 and O4, considering the best measured values. During O3 the losses were 12.5%, while we measured 3.2% for the new OMC.

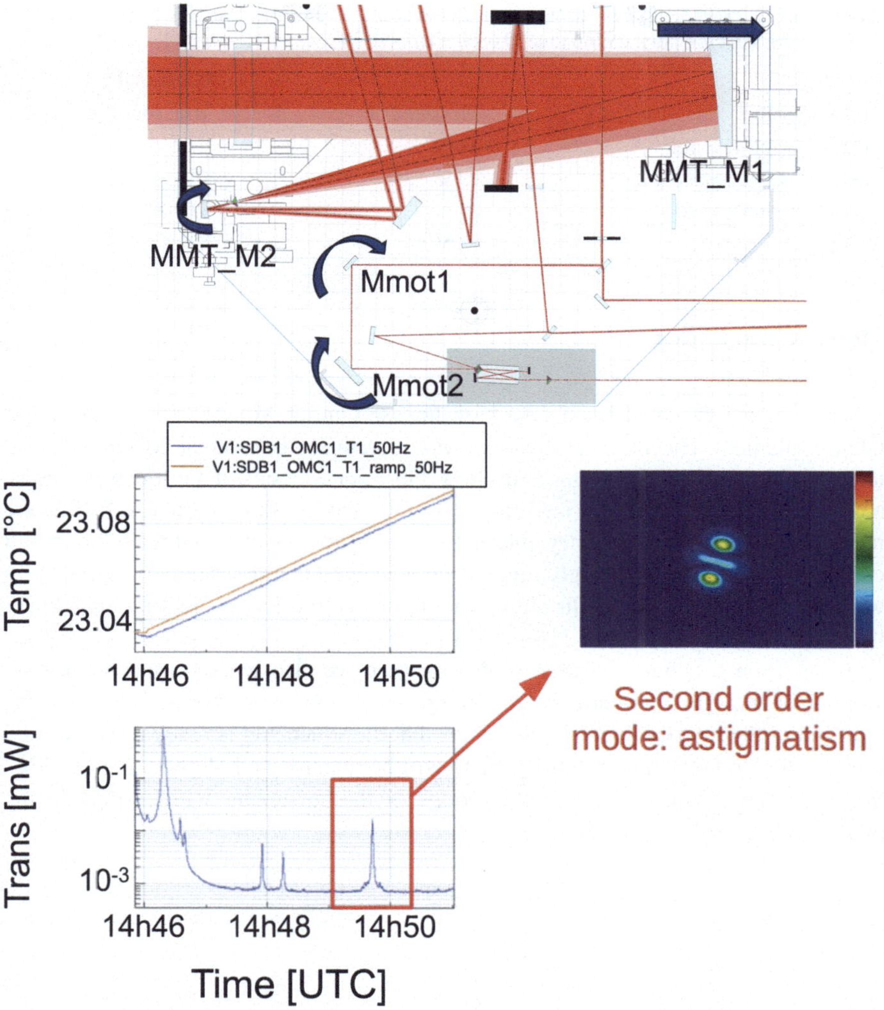

Fig. 10.17 Second method to improve the mode matching of the beam with the OMC: longitudinal displacement of the parabolic mirror MMT_M1, compensating with MMT_M2. The two mirrors before the OMC are used to compensate for any additional misalignment. The plots on the bottom show the power in transmission from the OMC during the scan in temperature. The second order mode shown on the camera is due to astigmatism

This results confirm that the installation of the new high finesse OMC cavity will allow to reduce detection losses as expected and therefore improve the squeezing performances.

Table 10.7 Measured losses comparison between the run O3 and O4. (*) alignment: 0.5% + MM: 0.64% + polarization tune: 0.08% = 1.2%

Losses sources	Measured O3	Measured O4 (pre-commissioning)
ITF—OMC1 matching	8%	1.2%*
OMC1—OMC2 matching	2.5%	0%
OMC internal losses	2%	2%
Total	12.5%	3.2%

10.6 Conclusion

In this chapter the need for a new high finesse Output Mode Cleaner cavity has been explained. The main motivation was to reduce the optical losses to avoid to spoil the level of squeezing and allow the improvement of the quantum noise reduction. We have characterized the new OMC cavity, showing that it fulfills the requirements (finesse, geometry, internal losses). Then we installed it and aligned it on the detection bench of Virgo. Finally, we measured the losses due to its matching with the ITF beam. The results shows that the detection losses are a factor 2 lower than in O3.

In addition to what is written in this chapter, we also worked on the commissioning of the OMC in dark fringe configuration and we performed the first DC read-out of the interferometer. This means that the squeezing is ready to be injected in the detector. Once the squeezing beam will be injected and aligned on the OMC, the longitudinal and angular control loops of the squeezing with respect to the interferometer will be engaged.

Part V
Outlook

Chapter 11
Prospects for Quantum Noise Reduction in Gravitational-Wave Detectors

The direct detection of gravitational waves has ushered in a new epoch for the observation of the universe, making it possible to investigate its hidden mysteries. In order to be able to see sources from ever more distant regions of the universe, gravitational wave detectors must be further improved. So far, a total of 90 coalescence events have been revealed, most of which are black hole collisions, two black hole-neutron stars coalescences and a two neutron stars coalescence. The latter allowed the beginning of the so-called multi-messenger astronomy. During the last O3 data take LIGO and Virgo detectors revealed one coalescence event per week. During O4, the sensitivity of Virgo will increase from 60 Mpc to about 100 Mpc, increasing the amount of observable events thanks to the reduction of several sources of noise. Among them, it is necessary to reduce quantum noise over the entire detection band by introducing frequency-dependent squeezed quantum vacuum states in the interferometer. The ellipse angle of the squeezed states needs to be rotated at the frequency where shot noise and radiation pressure noise have the same amplitude, i.e. 25 Hz for Virgo, to provide a broadband quantum noise reduction. During O3, the introduction of frequency-independent states improved Virgo's sensitivity at high frequencies by 3 dB. However, it was not possible to inject all the squeezing produced in order to avoid worsening the low-frequency sensitivity. To avoid this, it is necessary to introduce a filter cavity that rotates the squeezed states as a function of the frequency. This also makes it possible to increase low-frequency sensitivity if there are no other limiting noises. One of these noises is the scattered light from the optics on the suspended benches that re-couples with the interferometer's main beam. This noise source was studied, projecting its effect on Virgo's sensitivity curve.

In addition, the losses introduced on the squeezed states limit its performance: $(36.5 \pm 4.5)\%$ optical losses and (45 ± 40) mrad phase noise were measured. One of the largest contributions to optical losses comes from detection losses. In order to reduce these, the two output mode cleaner cavities were replaced with a single high finesse cavity.

E. Polini, *Broadband Quantum Noise Reduction in Advanced Virgo Plus*,
Springer Theses, https://doi.org/10.1007/978-3-031-95143-5_11

The frequency dependent squeezing technique was successfully implemented and commissioned in Virgo. The first demonstration of squeezed state rotation took place at a frequency of 361 Hz, but without reaching the level of shot noise. The aim is to produce frequency dependent squeezing states at 25 Hz in order to achieve a broadband quantum noise reduction in Virgo. After optimizing the system, 2 dB of frequency dependent squeezing was measured at low frequencies with rotation at 25 Hz. From this measurement, 14% optical losses (which, however, does not take into account losses in the interferometer) and 30 mrad phase noise were estimated. The noises affecting the squeezing needs to be reduced as much as possible to optimize its performances.

As far as scattered light is concerned, a complete simulation and mitigation study was conducted on all Virgo's suspended benches. In order to simulate the scattered light from the individual suspended optical elements more accurately, a back-scatterometer at LAPP was constructed with which we measured the BRDF of most of the critical elements. It was shown that the suspended detection benches SDB1 and SDB2 are the largest contributors to this noise. Furthermore, an active scattered light control loop was successfully implemented to reduce it on the squeezing system.

To reduce optical detection losses between O3 and O4, the two OMC cavities were replaced with a new high finesse cavity to allow filtering of the 6 MHz sidebands needed in O4 to control the interferometer. This eliminates losses due to mode mismatch between the two cavities, which accounted for 2.5% during O3. Mode matching and polarisation matching were also improved thanks to the motorized counter-weights and waveplates installed on the bench hosting the cavity. During commissioning, a mismatch of less than 1%, a mode mismatch of less than 1% and polarisation mistuning of less than 0.1% were achieved. In conclusion, the optical detection losses were reduced by half compared to O3. After optimizing the interferometer beam matching with the OMC, the DC readout of the interferometer was performed and the first sensitivity curve of the detector was extracted. The detector is therefore ready for the injection of the squeezing, since it has been obtained a pre-matching of 99% between the squeezing and the interferometer beam.

The next steps are to optimize the matching of squeezing with the interferometer, implementing longitudinal and angular control loops to keep the system stable during the data taking. Next, we will see the effect of frequency-dependent squeezing on Virgo's sensitivity curve. By producing 12 dB of squeezing, we aspire to see a 4.5 dB increase of the sensitivity at high frequency.

In order to benefit from the effect of low-frequency squeezing, we plan to continue the mitigation of stray light from suspended benches. I have shown that it is needed to continue the characterization of the optical elements (OMC cavity and Faraday Isolator TGG crystal) and take some further actions in order to reduce the impact of scattered light (install beam dumps in front of quadrant photodiodes). Finally, measurements of the scattered light coupling factors on Virgo's sensitivity curve are to be performed.

It is planned to improve the squeezing impact on the sensitivity curve of Virgo in O5 and post-O5. The level of squeezing produced in O5 and in post-O5 will be 12 dB, I.e. the same as in O4. Measured squeezing at high-frequency is expected to be 5.5 dB for O5 and 7.5 dB in post-O5. In order to improve the system, optical losses can be further reduced by placing the squeezer in a vacuum, so that an attenuation stage can be removed, i.e. removing a Faraday Isolator and replacing the Faraday Isolators with low-loss ones. To reduce the detection optical losses, it is planned to replace the monolithic OMC cavity with a four mirrors bow-tie ring cavity, as in LIGO, to eliminate optical losses by absorption in the fused silica.

Regarding the mitigation of scattered light in post O5, it has been considered to mount the input lenses of the suspended benches on motorized rotating supports in order to reduce specular reflections that re-couple with the Virgo main beam.

The work presented in this thesis and the prospects it opens up will allow Virgo's sensitivity to be improved, making it possible to see larger portions of the cosmos and to have a greater knowledge of the universe in which we live.

Appendix A
Alignment and Mode Matching of Gaussian Beams

If a laser beam is not perfectly aligned and mode matched to an optical cavity, it introduces optical losses on the squeezed field (see Sect. 3.4). The two cavities of interest in this work are the filter cavity (see Chap. 5) and the output mode cleaner cavity (see Chap. 10). The beam misalignment can be either transverse (as shown in Fig. A.1a) or angular (as in Fig. A.1b). Mode mismatch can be either due to a difference in beam waist size compared to the cavity (see Fig. A.1c) or due to a displacement of the waist position (see Fig. A.1d).

The solution of the scalar wave equation can be described by a set of complete and orthogonal functions, called *propagation modes*.

A.1 Hermite-Gaussian Modes

The modes that describe the misalignment of a Gaussian beam with respect to an optical cavity are called *Hermite-Gaussian (HG) modes* and are expressed with Cartesian coordinates. HG modes are the solutions of the scalar wave equation with electric field distribution given by

$$\begin{aligned} E_{nm}(x, y, z) \\ = E_0 \cdot \frac{w_0}{w(z)} \cdot H_n\left(\sqrt{2}\frac{x}{w(z)}\right) \cdot e^{-x^2/w(z)^2} \cdot H_m\left(\sqrt{2}\frac{y}{w(z)}\right) \cdot e^{-y^2/w(z)^2} \cdot \\ e^{-ikz} \cdot e^{-ik(x^2+y^2)/2R(z)} \cdot e^{i(1+n+m)\arctan(z/z_R)} \end{aligned} \tag{A.1}$$

for a field propagating along the z direction. $H_n(x)$ and $H_m(y)$ represents the Hermite polynomials with the non-negative integer indices n and m which determine the shape of the profile in the x and y directions, respectively. w_0 is the beam waist,

E. Polini, *Broadband Quantum Noise Reduction in Advanced Virgo Plus*,
Springer Theses, https://doi.org/10.1007/978-3-031-95143-5

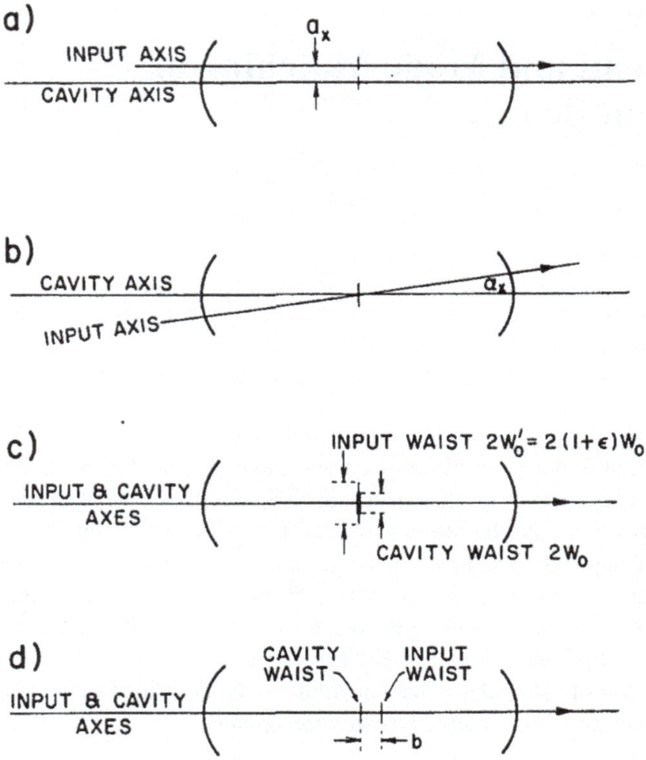

Fig. A.1 Possible misalignment and mismatch of the input axis with the cavity axis [119] (**a**) transverse displacement in the x direction; (**b**) tilt through an angle α_x; (**c**) waist size mismatch; (**d**) axial waist displacement

$w(z)$ is the beam size at the position z and z_R is the Rayleigh range (defined in Eqs. 5.1 and 5.2). The expression of the low order Hermite polynomials is

$$\begin{aligned} H_0(x) &= 1 \\ H_1(x) &= x \\ H_2(x) &= 4x^2 - 2 \\ H_3(x) &= 8x^3 - 12x. \end{aligned} \tag{A.2}$$

The intensity distribution of HG modes of order $N = n+m$ is shown in Fig. A.2.

The modes illustrated are visible, for example, in transmission from an optical cavity. For the beam in the cavity to be well aligned with it, it is necessary to act on the shift and/or tilt of the incident beam to reduce the order and power of the HG modes. In case of small misalignment of the input beam the first order mode dominates and the mode is vertical (horizontal) in case of vertical (horizontal)

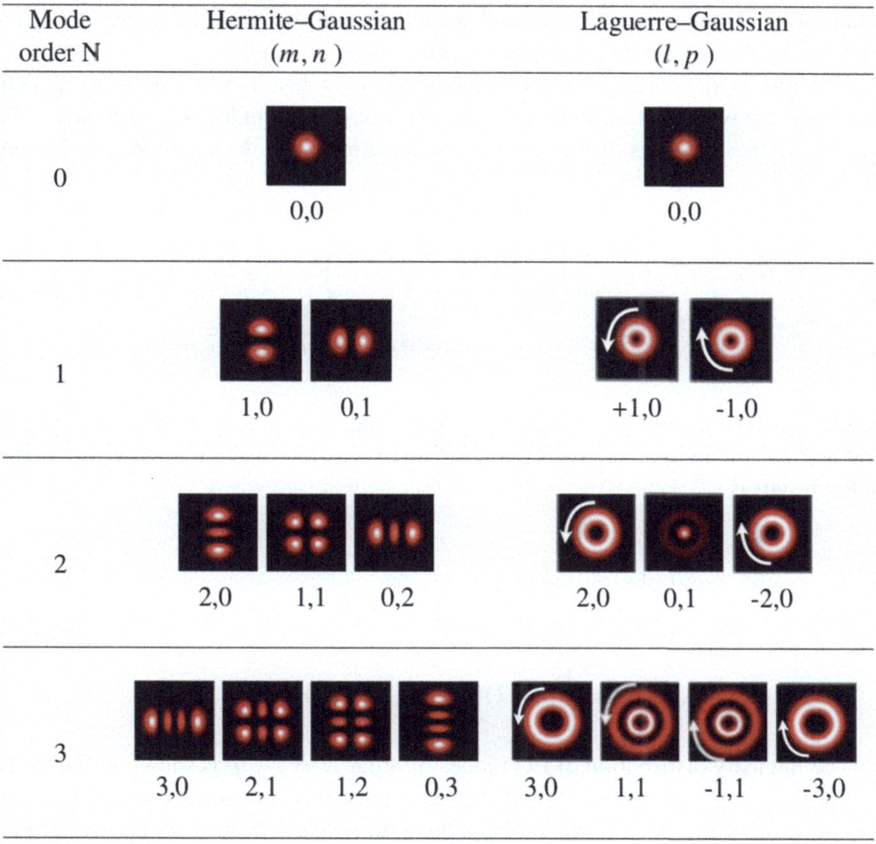

Fig. A.2 Hermite–Gaussian and Laguerre–Gausian mode indices and intensity profiles for mode order N ≤ 3. $N = n + m$ for HG modes and $N = l + 2p$ for LG modes. Figure from [120]

misalignment. The beam misalignment is given by the power ratio between the HG higher order modes and the fundamental mode TEM00 of the cavity. If the first order HG mode is smaller than the second order mode it points to an astigmatism of the input beam.

A.2 Laguerre-Gaussian Modes

If the beam is not perfectly matched to a cavity, take the example of Fig. A.1c with a waist mismatch, the input beam (w_0') can be decomposed on the basis of the cavity modes (with w_0) as

$$E_0' = \alpha \cdot E_0 + \epsilon \cdot E_{01} + \dots \tag{A.3}$$

where $\alpha \sim 1$ and $\epsilon << 1$ for small mistuning and the mode mismatching is ϵ^2 (defined by the amplitude of the second order mode of E_{01}).

The modes describing the mismatch mode of a beam in a cavity are called *Laguerre-Gaussian (LG) modes* and are expressed in cylindrical coordinates. LG modes are the solutions of the scalar wave equation with electric field distribution given by

$$E^l_p(r, \phi, z) = E_0 \cdot \frac{w_0}{w(z)} \cdot e^{-r^2/w(z)^2} \left(\frac{r\sqrt{2}}{w(z)}\right)^l L^l_p\left(\frac{2r^2}{w(z)^2}\right) \cdot \\ e^{-ikr^2/2R(z)} \cdot e^{i(1+2p+l)\zeta(z)} \cdot e^{-il\phi} \tag{A.4}$$

where L^l_p is a generalized Laguerre polynomial, with l and p called the radial and angular mode numbers respectively. The expression of the low order Laguerre polynomials is

$$\begin{aligned} L^l_0(x) &= 1 \\ L^l_1(x) &= l + 1 - x \\ L^l_2(x) &= \frac{1}{2}(l+1)(l+2) - (l+2)x + \frac{1}{2}x^2. \end{aligned} \tag{A.5}$$

The intensity distribution of LG modes of order $N = l + p$ is shown in Fig. A.2.

For the beam to be well matched with the cavity, it is necessary to act on the size and position of the waist of the incident beam to reduce the order and power of the LG modes. The beam mode matching is given by the power ratio on the LG higher order modes and the fundamental mode TEM00 of the cavity.

Appendix B
Ghost Beams Propagating Inside the Linking Tubes

The ghost beams generated on the optics at the ends of the SQB1 and SQB2 benches propagate inside the linking tubes and can strike their inner surface, producing scattered light that re-couples with the main beam. More details on this analysis can be found in [101].

B.1 SQB1-SQB2 Linking Tube

The pipe connecting SQB1 and SQB2 is 12 m long.

In the direction from SQB1 to SQB2, there are ghost beams generated by the SQB1_M22 and SQB1_M23 mirrors that cannot be stopped directly on the bench and therefore propagate into the pipe. Their propagation is shown in Fig. B.1 where the proposed position of the baffle is shown, i.e. 3 m away from the start of the tube. This is the position that allows the ghost beam to be sufficiently separated from the main beam and allows the baffle to catch the ghost beams before they hit the tube.

In the direction from SQB2 to SQB1, i.e. for the direction in reflection from the cavity, the ghost beams produced by SQB2_M1 and SQB2_M2 cannot be stopped on the bench as they overlap too much with the main beam. Actually, there is a risk of hitting the inner surface of the tube, which is why it was decided to install the baffles indicated by the vertical grey lines shown in Fig. B.2.

The baffle aperture has been chosen in order to avoid the clipping of the ghost beams 'a' and 'b' shown in Fig. 7.11. The aperture of the baffle should be

$$a_{baffle} = 4 \cdot w(z)_{GB_a,GB_b} + d_{MB-GB_a,GB_b} \tag{B.1}$$

with d_{MB-GB_a,GB_b} being the highest distance between the main beam and the ghost beam 'a' or 'b', and being $w(z)_{GB_a,GB_b}$ the highest beam size between the two ghost beams. The optimal radius of the baffle aperture is $a = 64$ mm.

E. Polini, *Broadband Quantum Noise Reduction in Advanced Virgo Plus*,
Springer Theses, https://doi.org/10.1007/978-3-031-95143-5

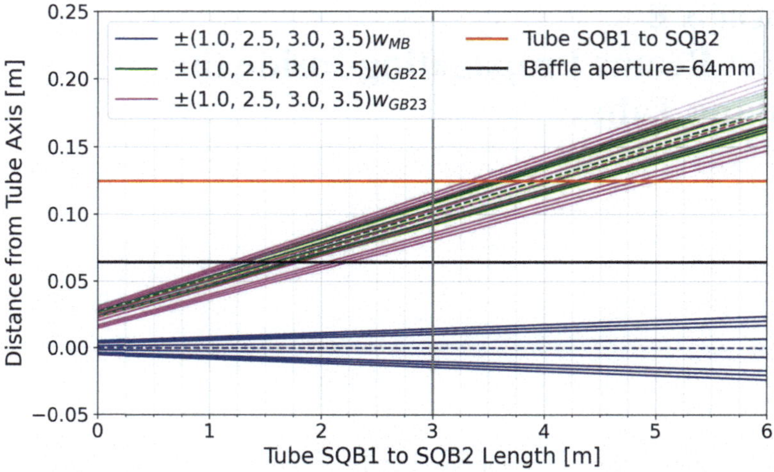

Fig. B.1 The main beam propagating from SQB1 to SQB2 inside the linking tube is represented in blue. The ghost beam generated by the mirror SQB1_M22 and propagating inside the tube is shown in green. The ghost beam generated by the mirror SQB1_M23 and propagating inside the tube is shown in pink. The optimal baffle position to properly dump the two ghost beams is at 3 m from the SQB1 edge, as shown by the vertical gray line (with a precision of few centimeters). The horizontal red lines represent the linking tube internal surface. The power of both ghost beams is 10^{-16} W

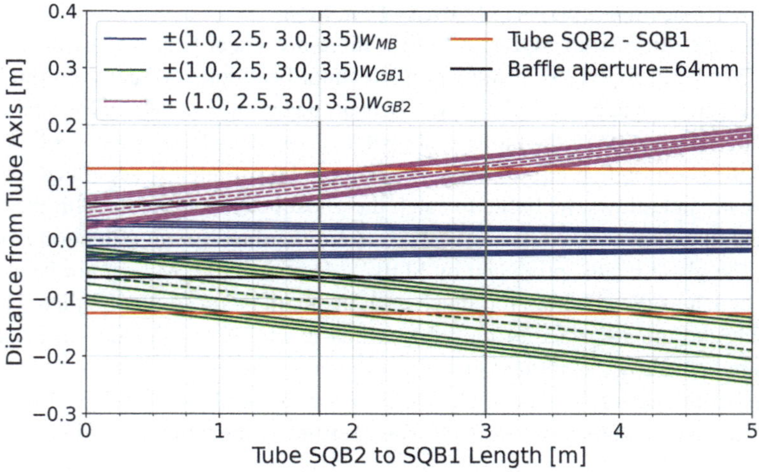

Fig. B.2 The main beam propagating from SQB2 to SQB1 inside the linking tube is represented in blue. The ghost beam generated by the mirror SQB2_M1 and propagating inside the tube, with a power of 10^{-16} W, is shown in green. The ghost beam generated by the mirror SQB2_M2 and propagating inside the tube, with a power of 10^{-11} W, is shown in pink. The optimal baffle positions to properly dump the two ghost beams are at 1.75 and 3 m from the SQB1 edge, as shown by the vertical gray lines (with a precision of few centimeters). The horizontal red lines represent the linking tube internal surface

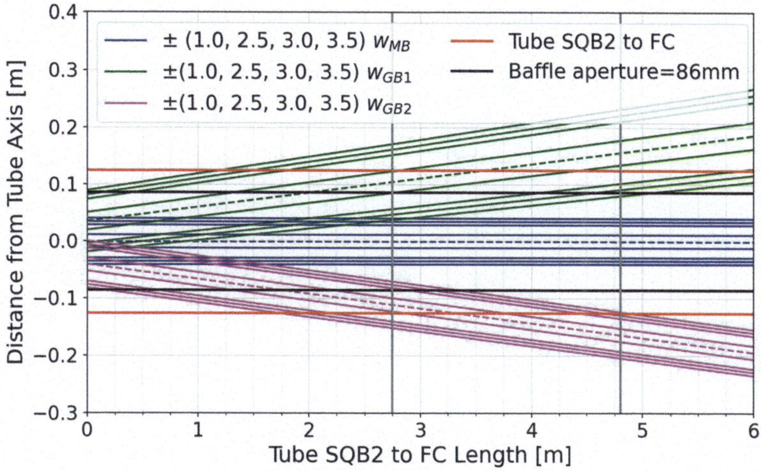

Fig. B.3 The main beam propagating from SQB2 to the filter cavity input mirror inside the linking tube is represented in blue. The ghost beam generated by the mirror SQB2_M1 and propagating inside the tube, with a power of 10^{-16} W, is shown in green. The ghost beam generated by the mirror SQB2_M2 and propagating inside the tube, with a power of 10^{-11} W, is shown in pink. The optimal baffle positions to properly dump the two ghost beams are at 2.75 and 4.80 m from the SQB2 edge, as shown by the vertical gray lines (with a precision of few centimeters). The horizontal red lines represent the linking tube internal surface

B.2 SQB2-FC Linking Tube

The pipe connecting SQB2 and the filter cavity input mirror tower is 40 m long.

The ghost beams generated by the SQB2_M1 and SQB2_M2 mirrors cannot be stopped directly on the bench and therefore propagate into the pipe. Their propagation is shown in Fig. B.3. To dump them two baffles are needed, one at 2.75 m from the beginning of the tube and the other at 4.80 m. Since the power of the ghost beam coming from SQB2_M2 is more powerful than the one coming from SQB2_M1, it should be stopped more carefully.

According to Eq. B.1, the optimal aperture radius of the baffle is $a = 86$ mm.

References

1. A. Einstein, Die Grundlage der allgemeinen Relativitätstheorie, 20 Mär 1916. The Foundation of the General Theory of Relativity, 20 Mar 1916 (1916)
2. M. Maggiore, *Gravitational Waves: Volume 1: Theory and Experiments*, vol. 1 (Oxford University Press, 2008)
3. A. Le Tiec, J. Novak, Theory of gravitational waves, in *An Overview of Gravitational Waves: Theory, Sources and Detection* (World Scientific, 2017), pp. 1–41
4. C.W. Misner, K.S. Thorne, J.A. Wheeler, Gravitation, vol. 891 (W. H. Freeman and Company, San Francisco, 1973)
5. M. Maggiore, *Gravitational Waves: Volume 2: Astrophysics and Cosmology* (Oxford University Press, Oxford, 2018)
6. B.P. Abbott, R. Abbott, T.D. Abbott, M.R. Abernathy, F. Acernese, K. Ackley, C. Adams, T. Adams, P. Addesso, R.X. Adhikari et al., First search for gravitational waves from known pulsars with Advanced LIGO. Astrophys. J. **839**(1), 12 (2017)
7. R. Abbott, T.D. Abbott, S. Abraham, F. Acernese, K. Ackley, A. Adams, C. Adams, R.X. Adhikari, V.B. Adya, C. Affeldt et al., All-sky search for continuous gravitational waves from isolated neutron stars in the early O3 LIGO data. Phys. Rev. D **104**(8), 082004 (2021)
8. R. Abbott, H. Abe, F. Acernese, K. Ackley, N. Adhikari, R.X. Adhikari, V.K. Adkins, V.B. Adya, C. Affeldt, D. Agarwal et al., All-sky search for continuous gravitational waves from isolated neutron stars using Advanced LIGO and Advanced Virgo O3 data. Preprint. arXiv:2201.00697 (2022)
9. R. Abbott, T.D. Abbott, S. Abraham, F. Acernese, K. Ackley, A. Adams, C. Adams, R.X. Adhikari, V.B. Adya, C. Affeldt et al., Upper limits on the isotropic gravitational-wave background from Advanced LIGO and Advanced Virgo's third observing run. Phys. Rev. D **104**(2), 022004 (2021)
10. R. Abbott, T.D. Abbott, S. Abraham, F. Acernese, K. Ackley, A. Adams, C. Adams, R.X. Adhikari, V.B. Adya, C. Affeldt et al., Search for anisotropic gravitational-wave backgrounds using data from Advanced LIGO and Advanced Virgo's first three observing runs. Phys. Rev. D **104**(2), 022005 (2021)
11. V. Ferrari, L. Gualtieri, P. Pani, *General Relativity and Its Applications: Black Holes, Compact Stars and Gravitational Waves* (CRC Press, 2020)
12. B.P. Abbott, R. Abbott, T.D. Abbott, M.R. Abernathy, F. Acernese, K. Ackley, C. Adams, T. Adams, P. Addesso, R.X. Adhikari et al., Observation of gravitational waves from a binary black hole merger. Phys. Rev. Lett. **116**(6), 061102 (2016)

E. Polini, *Broadband Quantum Noise Reduction in Advanced Virgo Plus*,
Springer Theses, https://doi.org/10.1007/978-3-031-95143-5

13. S.E. Gossan, P. Sutton, A. Stuver, M. Zanolin, K. Gill, C.D. Ott, Observing gravitational waves from core-collapse supernovae in the advanced detector era. Phys. Rev. D **93**(4), 042002 (2016)
14. B.P. Abbott, R. Abbott, T.D. Abbott, F. Acernese, K. Ackley, C. Adams, T. Adams, P. Addesso, R.X. Adhikari, V.B. Adya et al., GW170814: a three-detector observation of gravitational waves from a binary black hole coalescence. Phys. Rev. Lett. **119**(14), 141101 (2017)
15. B.P. Abbott, R. Abbott, T.D. Abbott, F. Acernese, K. Ackley, C. Adams, T. Adams, P. Addesso, R.X. Adhikari, V.B. Adya et al., GW170817: observation of gravitational waves from a binary neutron star inspiral. Phys. Rev. Lett. **119**(16), 161101 (2017)
16. M. Ajello, A. Allafort, M. Axelsson, L. Baldini, G. Barbiellini, M.G. Baring, D. Bastieri, R. Bellazzini, B. Berenji, E. Bissaldi et al., Fermi-LAT observations of LIGO/Virgo event GW170817. Astrophys. J. **861**(2), 85 (2018)
17. R. Abbott, T.D. Abbott, S. Abraham, F. Acernese, K. Ackley, A. Adams, C. Adams, R.X. Adhikari, V.B. Adya, C. Affeldt et al., Observation of gravitational waves from two neutron star–black hole coalescences. Astrophys. J. Lett. **915**(1), L5 (2021)
18. B.P. Abbott, R. Abbott, T.D. Abbott, S. Abraham, F. Acernese, K. Ackley, C. Adams, R.X. Adhikari, V.B. Adya, C. Affeldt et al., GWTC-1: a gravitational-wave transient catalog of compact binary mergers observed by LIGO and Virgo during the first and second observing runs. Phys. Rev. X **9**(3), 031040 (2019)
19. R. Abbott, T.D. Abbott, S. Abraham, F. Acernese, K. Ackley, A. Adams, C. Adams, R.X. Adhikari, V.B. Adya, C. Affeldt et al., GWTC-2: compact binary coalescences observed by LIGO and Virgo during the first half of the third observing run. Phys. Rev. X **11**(2), 021053 (2021)
20. R. Abbott, T.D. Abbott, F. Acernese, K. Ackley, C. Adams, N. Adhikari, R.X. Adhikari, V.B. Adya, C. Affeldt, D. Agarwal et al., GWTC-2.1: Deep extended catalog of compact binary coalescences observed by LIGO and Virgo during the first half of the third observing run. Preprint. arXiv:2108.01045 (2021)
21. R. Abbott, T.D. Abbott, F. Acernese, K. Ackley, C. Adams, N. Adhikari, R.X. Adhikari, V.B. Adya, C. Affeldt, D. Agarwal et al., GWTC-3: Compact binary coalescences observed by LIGO and Virgo during the second part of the third observing run. Preprint. arXiv:2111.03606 (2021)
22. B.P. Abbott, R. Abbott, T.D. Abbott, S. Abraham, F. Acernese, K. Ackley, C. Adams, V.B. Adya, C. Affeldt, M. Agathos et al., Prospects for observing and localizing gravitational-wave transients with Advanced LIGO, Advanced Virgo and KAGRA. Living Rev. Relativ. **23**(1), 1–69 (2020)
23. The LSC-Virgo-KAGRA Observational Science Working Groups, LSC-Virgo-KAGRA observational science white paper. LIGO-T2100289-v3, VIR-0753C-21, JGW-T2113074-v3 (2021)
24. J. Weber, Detection and generation of gravitational waves. Phys. Rev. **117**(1), 306 (1960)
25. A. Franzen, Componentlibrary (a vector graphics library for illustrations of optics experiments). *Creative Commons Attribution-NonCommercial*, 3 (2006)
26. A.E. Siegman, *Lasers* (University Science Books, 1986)
27. G. Vajente, Note on signal recycling I: Field equations. VIR-030B-08 (2008)
28. F. Ardhuin, E. Stutzmann, M. Schimmel, A. Mangeney, Ocean wave sources of seismic noise. J. Geophys. Res. Oceans **116**(C9) (2011). https://agupubs.onlinelibrary.wiley.com/doi/epdf/10.1029/2011JC006952
29. F. Acernese, P. Amico, N. Arnaud, D. Babusci, R. Barillé, F. Barone, L. Barsotti, M. Barsuglia, F. Beauville, M.A. Bizouard et al., Properties of seismic noise at the Virgo site. Classical Quantum Gravity **21**(5), S433 (2004)
30. T. Accadia, F. Acernese, P. Astone, G. Ballardin, F. Barone, M. Barsuglia, A. Basti, T.S. Bauer, M. Bebronne, M.G. Beker et al., A state observer for the Virgo inverted pendulum. Rev. Sci. Instrum. **82**(9), 094502 (2011)

31. S.A. Hughes, K.S. Thorne, Seismic gravity-gradient noise in interferometric gravitational-wave detectors. Phys. Rev. D **58**, 122002 (1998)
32. S. Sato, S. Miyoki, S. Telada, D. Tatsumi, A. Araya, M. Ohashi, Y. Totsuka, M. Fukushima, M.-K. Fujimoto, LISM Collaboration et al., Ultrastable performance of an underground-based laser interferometer observatory for gravitational waves. Phys. Rev. D **69**(10), 102005 (2004)
33. F. Badaracco, J. Harms, A. Bertolini, T. Bulik, I. Fiori, B. Idzkowski, A. Kutynia, K. Nikliborc, F. Paoletti, A. Paoli et al., Machine learning for gravitational-wave detection: surrogate Wiener filtering for the prediction and optimized cancellation of newtonian noise at Virgo. Classical Quantum Gravity **37**(19), 195016 (2020)
34. H.B. Callen, R.F. Greene, On a theorem of irreversible thermodynamics. Phys. Rev. **86**, 702–710 (1952)
35. K. Somiya, Detector configuration of KAGRA–the Japanese cryogenic gravitational-wave detector. Classical Quantum Gravity **29**(12), 124007 (2012)
36. V. Brisson, Virgo final design baffles (Virgo note 15/05/95). VIR-0393A-12 (1995)
37. M. Was, R. Gouaty, AdV+ noise review: DET – SDB1/SDB2/SPRB/SNEB/SWEB. VIR-1155B-19 (2019)
38. S.S.Y. Chua, B.J.J. Slagmolen, D.A. Shaddock, D.E. McClelland, Quantum squeezed light in gravitational-wave detectors. Classical Quantum Gravity **31**(18), 183001 (2014)
39. T. Accadia, F. Acernese, M. Alshourbagy, P. Amico, F. Antonucci, S. Aoudia, N. Arnaud, C. Arnault, K.G. Arun, P. Astone et al., Virgo: a laser interferometer to detect gravitational waves. J. Instrum. **7**(3), 3012 (2012)
40. F. Acernese, M. Agathos, K. Agatsuma, D. Aisa, N. Allemandou, A. Allocca, J. Amarni, P. Astone, G. Balestri, G. Ballardin et al., Advanced Virgo: a second-generation interferometric gravitational wave detector. Classical Quantum Gravity **32**(2), 024001 (2014)
41. J.C. Doyle, B.A. Francis, A.R. Tannenbaum, *Feedback Control Theory* (Courier Corporation, 2013)
42. E.D. Black, An introduction to Pound–Drever–Hall laser frequency stabilization. Am. J. Phys. **69**(1), 79–87 (2001)
43. M. Saleem, J. Rana, V. Gayathri, A. Vijaykumar, S. Goyal, S. Sachdev, J. Suresh, S. Sudhagar, A. Mukherjee, G. Gaur et al., The science case for LIGO-India. Classical Quantum Gravity **39**(2), 025004 (2021)
44. P. Amaro-Seoane, H. Audley, S. Babak, J. Baker, E. Barausse, P. Bender, E. Berti, P. Binetruy, M. Born, D. Bortoluzzi et al., Laser interferometer space antenna. Preprint. arXiv:1702.00786 (2017)
45. M. Punturo, M. Abernathy, F. Acernese, B. Allen, N. Andersson, K. Arun, F. Barone, B. Barr, M. Barsuglia, M. Beker et al., The Einstein Telescope: a third-generation gravitational wave observatory. Classical Quantum Gravity **27**(19), 194002 (2010)
46. D. Reitze, R.X. Adhikari, S. Ballmer, B. Barish, L. Barsotti, G. Billingsley, Duncan A. Brown, Y. Chen, D. Coyne, R. Eisenstein et al., Cosmic explorer: the US contribution to gravitational-wave astronomy beyond LIGO. Preprint. arXiv:1907.04833 (2019)
47. S.E. Gossan, E.D. Hall, S.M. Nissanke, Optimizing the third generation of gravitational-wave observatories for Galactic astrophysics. Astrophys. J. **926**(2), 231 (2022)
48. ET Steering Committee. ET design report update 2020. ET-0007B-20 (2020)
49. R.W. Boyd, *Nonlinear Optics* (Academic Press, 2020)
50. J. Bauchrowitz, T. Westphal, R. Schnabel, A graphical description of optical parametric generation of squeezed states of light. Am. J. Phys. **81**(10), 767–771 (2013)
51. M.O. Scully, M.S. Zubairy, *Quantum Optics* (American Association of Physics Teachers, 1999)
52. H.P. Yuen, V.W.S. Chan, Noise in homodyne and heterodyne detection. Opt. Lett. **8**(3), 177–179 (1983)
53. P. Kwee, J. Miller, T. Isogai, L. Barsotti, M. Evans, Decoherence and degradation of squeezed states in quantum filter cavities. Phys. Rev. D **90**(6), 062006 (2014)
54. J. Appel, D. Hoffman, E. Figueroa, A.I. Lvovsky, Electronic noise in optical homodyne tomography. Phys. Rev. A **75**(3), 035802 (2007)

55. S.S.Y. Chua, S. Dwyer, L. Barsotti, D. Sigg, R.M.S. Schofield, V.V. Frolov, K. Kawabe, M. Evans, G.D. Meadors, M. Factourovich et al., Impact of backscattered light in a squeezing-enhanced interferometric gravitational-wave detector. Classical Quantum Gravity **31**(3), 035017 (2014)
56. T. Aoki, G. Takahashi, A. Furusawa, Squeezing at 946nm with periodically poled ktiopo 4. Opt. Express **14**(15), 6930–6935 (2006)
57. S. Dwyer, L. Barsotti, S.S.Y. Chua, M. Evans, M. Factourovich, D. Gustafson, T. Isogai, K. Kawabe, A. Khalaidovski, P.K. Lam et al., Squeezed quadrature fluctuations in a gravitational wave detector using squeezed light. Opt. Express **21**(16), 19047–19060 (2013)
58. H.J. Kimble, Y. Levin, A.B. Matsko, K.S. Thorne, S.P. Vyatchanin, Conversion of conventional gravitational-wave interferometers into quantum nondemolition interferometers by modifying their input and/or output optics. Phys. Rev. D **65**(2), 022002 (2001)
59. C.M. Caves, B.L. Schumaker, New formalism for two-photon quantum optics. I. Quadrature phases and squeezed states. Phys. Rev. A **31**, 3068–3092 (1985)
60. B.L. Schumaker, C.M. Caves, New formalism for two-photon quantum optics. II. Mathematical foundation and compact notation. Phys. Rev. A **31**(5), 3093 (1985)
61. C.M. Caves, Quantum-mechanical noise in an interferometer. Phys. Rev. D **23**, 1693–1708 (1981)
62. A. Buonanno, Y. Chen, Quantum noise in second generation, signal-recycled laser interferometric gravitational-wave detectors. Phys. Rev. D **64**(4), 042006 (2001)
63. S.L. Danilishin, F.Y. Khalili, H. Miao, Advanced quantum techniques for future gravitational-wave detectors. Living Rev. Relativ. **22**(1), 1–89 (2019)
64. M. Evans, L. Barsotti, P. Kwee, J. Harms, H. Miao, Realistic filter cavities for advanced gravitational wave detectors. Phys. Rev. D **88**(2), 022002 (2013)
65. F. Acernese, M. Agathos, L. Aiello, A. Allocca, A. Amato, S. Ansoldi, S. Antier, M. Arène, N. Arnaud, S. Ascenzi et al., Increasing the astrophysical reach of the Advanced Virgo detector via the application of squeezed vacuum states of light. Phys. Rev. Lett. **123**(23), 231108 (2019)
66. F. Acernese, M. Agathos, L. Aiello, A. Ain, A. Allocca, A. Amato, S. Ansoldi, S. Antier, M. Arène, N. Arnaud et al., Quantum backaction on kg-scale mirrors: observation of radiation pressure noise in the advanced Virgo detector. Phys. Rev. Lett. **125**(13), 131101 (2020)
67. S. Chelkowski, H. Vahlbruch, B. Hage, A. Franzen, N. Lastzka, K. Danzmann, R. Schnabel, Experimental characterization of frequency-dependent squeezed light. Phys. Rev. A **71**(1), 013806 (2005)
68. E. Oelker, T. Isogai, J. Miller, M. Tse, L. Barsotti, N. Mavalvala, M. Evans, Audio-band frequency-dependent squeezing for gravitational-wave detectors. Phys. Rev. Lett. **116**(4), 041102 (2016)
69. L. McCuller, C. Whittle, D. Ganapathy, K. Komori, M. Tse, A. Fernandez-Galiana, L. Barsotti, P. Fritschel, M. MacInnis, F. Matichard et al., Frequency-dependent squeezing for advanced LIGO. Phys. Rev. Lett. **124**(17), 171102 (2020)
70. Y. Zhao, N. Aritomi, E. Capocasa, M. Leonardi, M. Eisenmann, Y. Guo, E. Polini, A. Tomura, K. Arai, Y. Aso et al., Frequency-dependent squeezed vacuum source for broadband quantum noise reduction in advanced gravitational-wave detectors. Phys. Rev. Lett. **124**(17), 171101 (2020)
71. L. Conti, A. Bertolini, S. Chua, Stray light noise in AdV from frequency dependent squeezing. VIR-0875B-17 (2017)
72. F.Y. Khalili, Optimal configurations of filter cavity in future gravitational-wave detectors. Phys. Rev. D **81**(12), 122002 (2010)
73. E. Capocasa, M. Barsuglia, J. Degallaix, L. Pinard, N. Straniero, R. Schnabel, K. Somiya, Y. Aso, D. Tatsumi, R. Flaminio, Estimation of losses in a 300 m filter cavity and quantum noise reduction in the KAGRA gravitational-wave detector. Phys. Rev. D **93**(8), 082004 (2016)
74. The Virgo Collaboration, Advanced Virgo Plus phase I - design report. VIR-0596A-19 (2019)

75. M. Mehmet, H. Vahlbruch, The squeezed light source for the advanced virgo detector in the observation run O3. Galaxies **8**(4), 79 (2020)
76. The AEI-AdV Squeezing Working Group, Integration of the AEI squeezer in the Advanced Virgo detector - Technical Design Report. VIR-0761B-17 (2017)
77. L. Naticchioni, S. Di Pace, M. Bawaj, FDS Homodyne Detector for AdV+. VIR-0698A-20 (2020)
78. K. McKenzie, M.B. Gray, P.K. Lam, D.E. McClelland, Technical limitations to homodyne detection at audio frequencies. Appl. Opt. **46**(17), 3389–3395 (2007)
79. L. Pinard, Input mirror characterization Advanced Virgo + filter cavity reference LMA: C20083 + C20087. VIR-1103B-20 (2020)
80. L. Pinard, End mirror characterization Advanced Virgo + filter cavity reference LMA: C20092 + C20087 (2021)
81. H. Tariq, A. Takamori, F. Vetrano, C. Wang, A. Bertolini, G. Calamai, R. De Salvo, A. Gennai, L. Holloway, G. Losurdo et al., The linear variable differential transformer (lvdt) position sensor for gravitational wave interferometer low-frequency controls. Nuclear Instrum. Methods Phys. Res. Sect. A Accel. Spectrom. Detect. Assoc. Equipment **489**(1–3), 570–576 (2002)
82. J.H. Voet, F. Schimmel, A. Bertolini, Coil driver/LVDT electronics for suspended benches - documentation. VIR-0548A-14 (2014)
83. The Virgo Collaboration, Advanced Virgo Technical Design Report. VIR-0128A-12 (2012)
84. R. Schilling, OptoCad – A Fortran 95 module for tracing Gaussian TEM00 beams through an opticalset-up. www.rzg.mpg.de/~ros/Optocad
85. S. Chelkowski, H. Vahlbruch, K. Danzmann, R. Schnabel, Coherent control of broadband vacuum squeezing. Phys. Rev. A **75**(4), 043814 (2007)
86. S. Soni, C. Austin, A. Effler, R.M.S. Schofield, G. González, V.V. Frolov, J.C. Driggers, A. Pele, A.L. Urban, G. Valdes et al., Reducing scattered light in LIGO's third observing run. Classical Quantum Gravity **38**(2), 025016 (2020)
87. A. Longo, S. Bianchi, W. Plastino, N. Arnaud, A. Chiummo, I. Fiori, B. Swinkels, M. Was, Scattered light noise characterisation at the virgo interferometer with tvf-EMD adaptive algorithm. Classical Quantum Gravity **37**(14), 145011 (2020)
88. M. Was, R. Gouaty, R. Bonnand, End benches scattered light modelling and subtraction in advanced virgo. Classical Quantum Gravity **38**(7), 075020 (2021)
89. B. Canuel, E. Genin, G. Vajente, J. Marque, Displacement noise from back scattering and specular reflection of input optics in advanced gravitational wave detectors. Opt. Express **21**(9), 10546–10562 (2013)
90. E.C. Fest, *Stray Light Analysis and Control*. Press Monographs (SPIE Press, Bellingham, 2013)
91. M. Was, E. Polini, High-angular-resolution interferometric backscatter meter. Opt. Lett. **47**(9), 2334–2337 (2022)
92. A. Bertolini on behalf of SBE, SBE technical noise. VIR-0964A-19 (2019)
93. J. Marque, E. Tournefier, B. Canuel, I. Fiori, Diffused light mitigation in Virgo and constraints for Virgo+ and AdV. VIR-0792A-09 (2009)
94. M. Evans, Optickle. LIGO-T070260-v1 (2007)
95. E. Genin B. Canuel, J. Marque, A high power, low diffusing and vacuum compatible beam dump for dumping B1s on SDB. VIR-0620A-10 (2010)
96. F.J. Raab, S.E. Whitcomb, Estimation of special optical properties of a triangular ring cavity. LIGO-T920004 (1992)
97. X. Chen, L. Ju, R. Flaminio, H. Lück, C. Zhao, D.G. Blair, Rayleigh scattering in fused silica samples for gravitational wave detectors. Opt. Commun. **284**(19), 4732–4737 (2011)
98. E. Genin, G. Pillant, A. Chiummo, R. Gouaty, Change request: Low-loss faraday isolators for squeezed light injection. VIR-0441A-17 (2017)

99. H. Yoshida, K. Tsubakimoto, Y. Fujimoto, K. Mikami, H. Fujita, N. Miyanaga, H. Nozawa, H. Yagi, T. Yanagitani, Y. Nagata, H. Kinoshita, Optical properties and faraday effect of ceramic terbium gallium garnet for a room temperature faraday rotator. Opt. Express **19**(16), 15181–15187 (2011)
100. E. Polini on behalf of DET sub system, DET: last installations on suspended benches - Virgo Week presentation. VIR-0826A-21 (2021)
101. E. Polini, R. Bonnand, E. Tournefier, Note on ghost beams study on frequency dependent squeezing system. VIR-0833B-20 (2020)
102. E. Polini, C. De Rossi, A. Menendez-Vazquez, Detection system ghost beam review (SDB1, SDB2). VIR-0187A-21 (2021)
103. E. Polini, C. De Rossi, A. Menendez-Vazquez, PSL/INJ ghost beam review (LB, EIB, SIB1, SIB2). VIR-0287A-22 (2022)
104. E. Polini, R. Bonnand, M. De Laurentis, E. Tournefier, Note on optimal aperture radius for diaphragms and baffles. VIR-0832B-20 (2020)
105. T.S. Khwaja, S.A. Reza, Low-cost gaussian beam profiling with circular irises and apertures. Appl. Opt. **58**(4), 1048–1056 (2019)
106. P. Belland, J.P. Crenn, Changes in the characteristics of a gaussian beam weakly diffracted by a circular aperture. Appl. Opt. **21**(3), 522–527 (1982)
107. J. Degallaix, OSCAR a Matlab based optical FFT code, in Journal of Physics: Conference Series, vol. 228 (IOP Publishing, 2010), p. 012021
108. F. Sorrentino, B1 beam size measurement. Logbook entry 41746 (2018)
109. E. Tournefier, Advanced Virgo output mode cleaner: specifications. VIR-0071A-08 (2008)
110. M. Ducrot, Etude des cavités optiques de filtrage de sortie du détecteur d'ondes gravitationnelles Advanced Virgo. PhD thesis, Université Grenoble Alpes (ComUE), 2016
111. J. Marque, G. Vajente, Output mode cleaner length noise. VIR-0414B-12 (2012)
112. R. Gouaty et al., Specifications for the coating of OMC cavities of finesse = 1000. VIR-0523A-22 (2022)
113. R. Gouaty, M. Was, AdV+ noise review: DET - OMC length noise. VIR-0384A-20 (2020)
114. M. Was, J. Casanueva, F. Nocera, M. Mantovani, Requirement on relative amplitude modulation noise for O4. VIR-1225B-19 (2019)
115. N. Straniero, J. Degallaix, R. Flaminio, L. Pinard, G. Cagnoli, Realistic loss estimation due to the mirror surfaces in a 10 meters-long high finesse Fabry-Perot filter-cavity. Opt. Express **23**(16), 21455–21476 (2015)
116. R. Gouaty, M. Was, Specifications for the polishing of the Advanced Virgo "mode cleaner" cavities. VIR-0535A-21 (2021)
117. Personal communication between L. Pinard and R. Gouaty (2022)
118. P.A. Casula, L. Digallo, R. Gouaty, F. Marion, B. Mours, L. Rolland, Advanced Virgo output mode cleaner: refinement of the design and polishing specifications. VIR-0319A-12 (2012)
119. D.Z. Anderson, Alignment of resonant optical cavities. Appl. Opt. **23**(17), 2944–2949 (1984)
120. S. Restuccia, D. Giovannini, G. Gibson, M. Padgett, Comparing the information capacity of Laguerre–Gaussian and Hermite–Gaussian modal sets in a finite-aperture system. Opt. Express **24**(24), 27127–27136 (2016)

The manufacturer's authorised representative in the EU is Springer Nature Customer Service Centre GmbH, Europaplatz 3, 69115 Heidelberg, Germany. If you have any concerns regarding our products, please contact ProductSafety@springernature.com

Printed and bound by CPI Group (UK) Ltd, Croydon, CR0 4YY
30/09/2026
02225564-0007